Building and Maintaining Award-Winning ACS Student Member Chapters Volume 3

ACS SYMPOSIUM SERIES **1278**

Building and Maintaining Award-Winning ACS Student Member Chapters Volume 3

Matthew J. Mio, Editor
University of Detroit Mercy
Detroit, Michigan

Mark A. Benvenuto, Editor
University of Detroit Mercy
Detroit, Michigan

Sponsored by the
ACS Division of Chemical Education

American Chemical Society, Washington, DC

Distributed in print by Oxford University Press

Library of Congress Cataloging-in-Publication Data

Names: Mio, Matthew J. (Matthew John), 1974- editor. | Benvenuto, Mark A. (Mark Anthony), editor. | American Chemical Society. Society Committee on Education.
Title: Building and maintaining award-winning ACS student member chapters / Matthew J. Mio, editor, University of Detroit Mercy, Detroit, Michigan, Mark A. Benvenuto, editor, University of Detroit Mercy, Detroit, Michigan ; sponsored by the ACS Society Committee on Education.
Description: Washington, DC : American Chemical Society, [2016] | Series: ACS symposium series ; 1229, 1230, 1278 | Includes bibliographical references and index.
Identifiers: LCCN 2016043973 (print) | LCCN 2016044818 (ebook) | ISBN 9780841231696 (alk. paper : v. 1) | ISBN 9780841231719 (alk. paper : v. 2) | ISBN 9780841233850 (alk. paper : v. 3) | ISBN 9780841233843 ()
Subjects: LCSH: American Chemical Society--Membership. | Chemistry--Study and teaching (Higher)
Classification: LCC QD1 .B81495 (print) | LCC QD1 (ebook) | DDC 540.6/073--dc23
LC record available at https://lccn.loc.gov/2016043973

The paper used in this publication meets the minimum requirements of American National Standard for Information Sciences—Permanence of Paper for Printed Library Materials, ANSI Z39.48n1984.

Foreword

The ACS Symposium Series was first published in 1974 to provide a mechanism for publishing symposia quickly in book form. The purpose of the series is to publish timely, comprehensive books developed from the ACS sponsored symposia based on current scientific research. Occasionally, books are developed from symposia sponsored by other organizations when the topic is of keen interest to the chemistry audience.

Before agreeing to publish a book, the proposed table of contents is reviewed for appropriate and comprehensive coverage and for interest to the audience. Some papers may be excluded to better focus the book; others may be added to provide comprehensiveness. When appropriate, overview or introductory chapters are added. Drafts of chapters are peer-reviewed prior to final acceptance or rejection, and manuscripts are prepared in camera-ready format.

As a rule, only original research papers and original review papers are included in the volumes. Verbatim reproductions of previous published papers are not accepted.

ACS Books Department

Contents

Indexes

Preface

Perfection is not attainable,
but if we chase perfection
we can catch excellence.

– Vince Lombardi

Can one correlate the philosophical musings of one of the most famous football coaches in history with the best ACS Student Member Chapters? Yes! The link is in the excellence. Award-winning Student Member Chapters – several leaders of which have been kind enough to write a chapter in this volume – all have caught excellence in one or more facets of what they do.

We began this journey to capture the best of our Student Member Chapters back in 2015 when we asked some of the best and most active organizations' leadership to put into words what they did that puts them at the top. Those first thoughts came to fruition in print in 2016 with the twin ACS Books volumes 1229 and 1230, titled: "Building and Maintaining Award-Winning ACS Student Member Chapters, Volume 1: Holistic Viewpoints," and "Volume 2: Specific Program Areas." These two volumes endeavored to answer some basic questions for anyone looking to build a first-class Student Chapter of their own. What makes a chapter a winner? Is it the advisor, the students, the opportunities and events, or something more? We realized there is no one, specific answer to such questions, but found a wealth of information in what our chapter authors reported.

But there are more voices in this wonderful chorus, voices of leaders who have great ideas, and who have figured out ways to make the fascination of chemistry communicable to our students and the general public. The volume that you are holding represents some excellent input as to what makes a chapter award-winning, and what keeps its excellence sustainable. The work herein represents a 3rd volume in what has become a larger set, and we hope it can be a huge help, resource, and guide to any reader who wants to improve their own ACS Student Member Chapter.

As per our first two volumes, several of the chapters here have been written or co-written by Student Member Chapter advisors. These individuals can have a long-standing effect on a Chapter simply because an advisor is usually a faculty member who has built her or his career at an institution. Such individuals, when they pass on a positive attitude, influence ACS Student Members in far-reaching and long-lasting ways. The same holds true in Volume 3. In addition, the energy and attitude of the student leadership can be sensed coursing through the words in this volume, as well. To massage a very old axiom: the total of students plus advisors is definitely much more than the sum of the parts.

Vince Lombardi, one of the winningest football coaches in history, compiled an overall record of $105 - 35 - 6$ throughout his career. He undoubtedly had the same kind of energy that our Student Member Chapter advisors and student leaders have, and spread it to his players, much like our authors spread theirs (curiously, he also apparently taught both chemistry and Latin a bit before taking off as a coach). Importantly, Lombardi's record shows he did not attain perfection, but that he did catch that excellence. We bring you a group of chapters that hope to do the same.

Matthew J. Mio
Department of Chemistry and Biochemistry
University of Detroit Mercy
4001 West McNichols Road
Detroit, MI 48221-3038

Mark A. Benvenuto
Department of Chemistry and Biochemistry
University of Detroit Mercy
4001 West McNichols Road
Detroit, MI 48221-3038

Learning through Teaching and Growth through Service: Integrating the Chemistry Club Experience Throughout the Academic Major

Michelle A. Boucher,* Alyssa C. Thomas, and Curtis R. Pulliam

Utica College, 1600 Burrstone Rd., Utica, New York 13502, United States
*E-mail: mboucher@utica.edu

The chemistry major experience at Utica College is built around the question of what it means to be a chemist: how to think like chemists, work like chemists, and communicate like chemists. The more our students acted like chemists, the better off they were at visualizing what they could become, which we found helped us with retention, graduation, and post-graduate outcomes within the major. This realization led to an approach to our chapter and major that has been highly successful both for our department and the health of our student chapter. We now nurture our student ACS chapter in our major courses and integrate club activities throughout the major experience. Our chapter conducts activities intended to promote professional development, provide outreach to the larger community, and allow for quality group interactions in social settings. Integrating these events in a thoughtful way throughout the 4 years of the major experience has allowed student members to gain organizational and leadership skills, establish a network of peer mentorship, and gain confidence as emerging professionals.

School, Department, and Chapter Demographics

Utica College is a small, comprehensive, private college in central New York. It was founded by Syracuse University in 1946 and became independent in

1995 (*1*). The total number of students, undergraduate and graduate combined, is around 5000. While Utica College offers some masters and doctoral degrees, the Department of Chemistry and Biochemistry exclusively offers an undergraduate major. It is an ACS-approved program, with B.S. degrees in chemistry and biochemistry including ACS-certified degrees.

Our department is a close-knit group of six faculty members and 45–50 chemistry and biochemistry majors across all 4 years. Our departmental student demographics include a large number of first generation college students, 10%–20% students from underrepresented minorities, and about 40% of students who commute. We start with a freshman class with about a 2:1 ratio of biochemistry to chemistry majors with most of the majors identifying as premedical or other prehealth professional majors (pharmacy, dentistry, etc.). This distribution changes over time to a ratio closer to 1:1 and with a distribution of approximately 1:1:1 graduate school:health professional graduate schools (including medical school):direct entry into the workforce post graduation.

The Utica College Student Chapter of ACS was founded in 1976 and was reactivated in 2006 after a period of inactivity. Since our reactivation we have been a busy and active chapter with a strong departmental and school-wide presence. We hold multiple outreach events each semester, have at least one significant professional development activity each semester, and have consistently maintained high levels of recognition by ACS (two Honorable Mentions and nine Commendable ratings in the last 11 years).

Like many schools of our size and with our demographics, Utica College faces challenges with student retention both in the major and at the college. As a school with a large first-generation college-student population, many of our students come to us without an understanding of how much college and high school differ. Many come underprepared academically, needing to take lower-level math and English courses before they can start the math courses required by their major. Utica College, like many other schools, has made an effort to have students quickly identify with a major as a method to promote a sense of belonging and help them stay in their programs (*2*).

Learning To Be a Chemist

The chemistry major experience at Utica College was built around the question of what it means to be a chemist: Our classes and laboratories have the goals of teaching majors how to think like chemists, work like chemists, and communicate like chemists. Students need to be able to approach problems from a molecular worldview, to use instruments and techniques unique to the chemical sciences, and learn how to communicate their understanding of the molecular world to scientists and nonscientists alike. We want students to feel like chemical professionals, using the intellectual and physical tools of chemists from model kits through advanced instrumentation, starting with general chemistry and continuing through to graduation. From individual laboratory experiences through the creation of new classes and complete reworkings of traditional foundational chemical experiences, we have looked for ways to empower students to feel confident as chemical scientists (*3, 4*).

It became clear to us as a department that one of the biggest challenges our students faced was how to translate the idea of an academic major into plans for a future career. Just because we were teaching students how to think, work, and communicate like chemists did not mean that they understood all the different things chemists can do professionally, nor did it mean they would be confident in seeking a job that they had never seen in action. Our majors are typically premedical or prepharmacy as first-year students and they plan those career paths because those are the ones they recognize in their daily life. An informal survey of first-year majors in the department taken over the last 2 years indicated that about 75% of the majors had no idea that one could find a job as a bench chemist after graduation. Although we do not have perception data from before 2 years ago, anecdotally it seems valid that most majors in our recent past did not realize that chemical jobs could be options for them, nor did they know about graduate school in the chemical sciences. Historically, many of those majors became lost in their sophomore year once they determined that the prehealth pathway was either not what they wanted or no longer viable.

We began to see that those students who were active in the chemistry club, who had exposure to other career possibilities such as those highlighted in ACS publications we shared in the club, including one of the many "Careers in Chemistry" issues of *inChemistry* (*5*) and who had hands-on experiences outside the classroom, were more confident and were more likely to stay enrolled in the school and in the major. Many of our students were the first of their families to pursue undergraduate degrees and suffered from imposter syndrome (*6*). While we knew they were solid students and future chemists, they could not always see their own accomplishments and potential. Students expressed concerns that they couldn't possibly hope to be real scientists since they weren't smart enough or did not start "early" enough (meaning declaring intent in high school) or did not go to the right high school/grade school. By being part of our student chapter these students became much more comfortable and realistic in their assessment of their abilities; these juniors and seniors were impressive young scientists and they were now able to realize that fact. The more our students acted like chemists, being recognized in the scientific and general community as emerging professionals, the better off they were at visualizing what they could become.

This realization led to an approach to our chapter and major that has been highly successful for our department in terms of major retention, student success within the major and post-graduation, and in the health of our student chapter. In terms of graduation, our department has looked at graduation rates from incoming cohorts starting in 2009, when we began deliberately implementing this integrated approach, through 2012. We have seen an average 4-year graduation rate of 74% in our department for this time period, in comparison to an institutional graduation rate that averages around 45%. In the past 5 years, over 90% of our majors have gone on directly to their postgraduate programs of choice or found employment in a chemical field within 6 months of graduation. The health of our student chapter is robust, with over half of the majors being active members and ACS members at any given time.

We now nurture our student chapter of ACS in our courses for majors and integrate club activities throughout their experience. These activities (including outreach activities as a cornerstone) have built a chemical community within our major, allowing students to feel like and eventually become chemical professionals while also feeling that the department has become their home.

Chapter Events Integrated through the Major Experience

Our chapter conducts activities intended to promote professional development, provide outreach to the larger community, and allow for quality group interactions in social settings. Integrating these events in a thoughtful way throughout the 4 years of the major's experience has allowed student members to gain organizational and leadership skills, establish a network of peer mentorship, and gain confidence as emerging professionals.

Freshman Year

At Utica College in their first semester freshman students are placed into a first-year seminar course focused on introduction to college life. Students are typically grouped by majors in this course, and the chemistry and biochemistry section has always been taught by our department faculty. This course provides an opportunity to introduce what it means to be a chemist as well as explore major and career opportunities. A few years ago we started to schedule the first-year seminar at the same time as our senior seminar course so that we could bring together all the students for a variety of activities. One of the first activities that occurs early on in the semester is a visit from our student chapter executive board and members of senior seminar. They engage in icebreaker activities to get to know each other better and at the end of class the freshmen are invited to attend the first chemistry club meeting. This establishes a nice rapport between the freshmen and upperclassmen. The freshmen recognize not just the faculty but also our chapter members. They realize their teaching assistant for lab is also the student chapter secretary or president. Building these connections early is crucial for recruiting new members and retaining them over the next 4 years.

As a follow-up to that initial meeting, our student chapter puts together a social event to welcome the freshmen and new members. Our most popular one is a movie night on campus complete with comfort food—our mac'n'cheese bar is always a big hit. This gives the freshmen a chance to connect with the upperclassmen and faculty in an informal setting and establish strong chapter connections for the coming year.

Mentoring is another facet to the integration of majors in our department and our student chapter, which we promote during the freshman seminar course. We have the first-year students interview two upperclassmen to find out which chemistry classes were their favorites, what academic advice they wish they knew earlier, and what career goals they have. After the interviews, chapter members volunteer as mentors and are paired with first-year students with similar interests. The mentors bring the new students to chapter meetings, take them under their

4

wing during the planning and execution of outreach activities, and sometimes act as tutors to help them through tough courses. While officially these pairings are for the first year, the established relationships often last well past graduation.

During their second semester, the hands-on mentoring continues and the students are enrolled in CHE 213—Introduction to Research Methods. We specifically designed this course to introduce key concepts of the major experience: writing a lab report; ownership of samples and data; spectral analysis; and use of instrumentation (3). The students make aspirin in their general chemistry II course, analyze the impure sample in CHE 213, and provide evidence in their lab report as to whether they successfully synthesized aspirin—or not. We also invite each freshman to become an ACS student member as part of the course assignments and many will maintain their membership throughout their undergraduate career.

The chapter members help out when it comes to analyzing spectra. After the theories and concepts for IR, ^{1}H and ^{13}C NMR, and GC-MS are discussed in class, members volunteer to help with spectral analysis and walk the students through the process. Along with the instructor, they ask the students questions, look over peak assignments, and help with determining the identity of sample peaks versus impurities or starting material. These exercises help to instill confidence in interpreting data as well as the important concept that chemists typically work with other chemists. We ask each other for help and work collaboratively to solve problems, especially in a research setting. It is great to see upperclassmen and freshmen in the hallway discussing the proton resonance assignments in their NMR spectra.

Another important aspect of research is communicating results. Each spring, majors who will be attending conferences present in a chapter-run poster and oral presentation session for CHE 213 students. The CHE 213 students listen to the presentations, fill out evaluations, and are required to ask at least one question of the presenters. Freshmen are exposed to the different kinds of research projects going on in the department through peer interactions, upperclassmen gain valuable presentation practice in a nonthreatening environment, and chapter officers get experience running a mini meeting. Just as importantly, the CHE 213 students see their peers—Utica College students just like them—preparing to present their own research in professional settings to the greater scientific community. Their own chapter members, students just like them, are doing valuable chemistry and communicating it formally to other chemists. They see that they can do it, too.

Sophomore and Junior Years

Sophomore and junior years are developmental years during which our students are establishing themselves in their major and taking leadership roles in the chapter. If interested, students begin research with us at this point. Starting in their second year, students are encouraged by the department and the chapter to become laboratory assistants. Even though they have only had general chemistry up to this point, they are certainly able to help out in the general chemistry laboratory while studying organic chemistry. The chapter holds start-of-semester laboratory training events to encourage members to step outside their comfort zones and try helping out in the lab. When organic laboratory becomes frustrating it is good to be able to assist in a general chemistry laboratory; what seemed so difficult the year before seems easy now. Students really perceive how far they have come in just one year. Students in biochemistry and analytical chemistry come back to help in the organic laboratory and find again that they actually learned something and understand organic well enough to help other students.

We strive to promote a culture of chemical safety in our department as well as in our chapter activities. We firmly believe the more interest the students take in making appropriate safety choices on their own, the better chemists they will be. This led to the creation of the goggle ceremony. Up to this point, the students have been required to wear splash-proof goggles in the general and organic laboratories. During the first meeting of the majors' analytical lab, typically taken the fall of junior year, each student receives their first pair of sidearm goggles in a ceremony run by our student chapter and chapter advisers. We open the ceremony by talking about the importance of safety and choosing the right equipment for the task. This includes which type of safety goggle is most appropriate for each experiment including running instruments for analysis. Then we offer them an opportunity to discuss when each type of goggle is best to wear given the situation. Each student is called to the front of the lab to receive their sidearm goggles and shake the faculty's hands. They exit the lab to find additional faculty and other chapter members lining the hallway to cheer them on receiving upperclassmen status. They are now leaders in the chapter and in the major. We also take a group picture to commemorate the occasion.

Senior Year

Seniors are encouraged to take advantage of activities in the chapter to practice and learn leadership techniques. We specifically integrate chapter activities into our Senior Seminar CHE 495 course. Our CHE 495 course is where we work on the questions of what it means to be a chemist and how to prepare to enter the chemical community as a professional since, no matter what our students end up doing with their chemistry degrees, they have the right to call themselves chemists and biochemists. By the time our students are seniors we find that many have switched from biochemistry to chemistry and we are closer to a 1:1 ratio between the two disciplines. This is often a result of shifts from premedical pathways to other pathways, since students who were initially premedical often switch to plans of graduate school or direct entry into the workforce.

Senior seminar, a one-credit seminar, is a two-semester sequence with the fall being built in part around National Chemistry Week (NCW) and the spring being built in part around Chemists Celebrate Earth Week (CCEW).

Every semester has in it the required "text" of becoming an ACS member, since we use *C&E News* in the course. We start by reading *inChemistry* and *C&E News* articles and progress to searching, reading, and discussing the chemical literature. We also work on our professional writing such as resumes and personal statements; communicating like a chemist also means communicating who you are as a chemist. Chapter resources such as Resume Workshops are useful to students in the course. Each semester has a theme around which we build the topic we will study, and usually that theme is tied to either the NCW or the CCEW themes and both overlap with and inform what the chapter is doing. Discussing these "chemistry holidays" in class and integrating them into the materials not only helps keep the discussions fresh, but also inspires our seniors to plan chapter events surrounding these themes.

The fall semester of senior seminar has an outreach component in which we design and perform some community event following the theme of NCW as part of the course. In the spring, we focus our readings and discussions around the CCEW topic or themes surrounding the topic. While an outreach component is not required, typically students are inspired to integrate something that they have learned into a chemistry activity.

Senior seminar is also a place where it is natural to brainstorm professional development opportunities; many events such as our trips to hear local speakers, or the visits of invited speakers to campus, or travel to regional meetings such as the North East Regional Meeting (NERM), had their beginnings in senior seminar. Not every senior student is an active chapter member; typically a senior seminar class consists of 10–12 students only half of whom are active chapter members. This way, however, every student gains the benefits of what the chapter has to offer. Senior seminar is not an executive committee meeting of the chapter, it is a place where seniors become the leaders of our major as they learn what it means to be a chemical professional. Integration of the chapter into these activities is simply a method to help bring about that shift from undergraduate student to chemist.

Sample Chapter Activities

While there would seem to be a clear separation among outreach activities, professional development activities, and social activities, we have found that for a department and club as small and close-knit as we are, almost every event has a social aspect to it. Outreach events end with a trip for ice cream to celebrate; trips to regional meetings often involve a pizza run and a sightseeing tour. Even activities that seem on the surface to be completely social, like a bowling night with the club or a movie night on campus, are opportunities for peer-to-peer mentoring or an impromptu review session of an advanced course between bowling frames. Since our chapter has such a strong history of peer-to-peer mentorship, and since it is encouraged throughout the major experience, our students are savvy enough to turn any event into an opportunity to help each other out.

Outreach Activities

Our outreach activities are built around the idea that while outreach is great for the community, it is even better for our majors who learn how to become chemists through doing chemistry. While our college does not have a formal service-learning component, the function of the chemical outreach we perform provides a similar richness of experience. Each type of outreach event serves a different purpose for our majors and provides a different experience to them as they organize it and complete the event.

We do a number of activities centered around demonstrating chemical concepts or providing hands-on chemistry experiences. The formation of thoughtfully and deliberately designed safety protocols—for our students, for the community being served, and for both the physical space and the environment in general—is the first consideration for all of these events. Learning about chemical safety and communicating safety issues and concerns are some of the most important goals that we have for our students during the planning stage of any event.

We have developed a stock set of demonstrations and hands-on experiences that are well tested, come from reliable sources such as the ACS education website (*7*) or *Chemical Demonstrations* by Shakhashiri (*8*), and that our chapter has a history of running successfully. For each event we hold a safety briefing in advance of the outreach day, ideally run by chapter officers under the guidance of a faculty adviser. Here is where chemical safety concerns are discussed, such as appropriate personal protective equipment, hazards of the chemicals involved, safe chemical handling and disposal, and common potential mistakes/pitfalls of the experiments involved. An event would be cancelled if there were not enough time to have this pre-event run-through and safety briefing, or if there were not enough personal protective equipment available (such as loaner lab goggles).

Having student members take leadership roles in teaching safety has contributed to a culture of chemical safety within the department. Students are much more likely to keep to strict safety guidelines if they are encouraged to do so through peer-to-peer interaction. Student leaders feel more self-confident after successfully running both the safety pre-event and the event itself. Students who take leadership roles in teaching chemical safety are more likely to feel confident enough to become laboratory assistants for us; these laboratory assistants, in turn, are trained to think of safety first and thus help promote our commitment to safety. The experiences gained through these processes have in turn helped our graduates gain admission to and achieve success in graduate school.

Role-Playing as Chemists: Students Visit Our School

Our most frequent outreach efforts are on-campus hands-on events. In these events visiting students come to our laboratories and run an experiment or a series of experiments with us. They use standard chemical glassware, work on our benchtops or in our hoods, and analyze their results using our instrumentation. Groups served recently include entire classes/grades from local grade and high schools, Boy Scout troops, Girl Scout troops, Young Scholars students visiting campus as part of mini college experiences, and students visiting our Collegiate Science and Technology Entry Program (CSTEP), such as groups from the Oneida Indian Nation. We have served groups as small as eight students and as large as 40 and are planning to host even bigger events in the future.

For younger students, kindergarten through about third grade, we stick to well-established hands-on activities that are clearly linked to one chemical concept; UV active beads and light, shaving cream with food coloring on paper for polarity, measuring acidity with baking soda or pH paper made from red cabbage juice. These events work best for us when we have a number of stations, each with a dedicated chapter student running the station, and small groups of visiting students with one dedicated chapter student as escort (typically in 2:1 up to 4:1 ratios) as they move from station to station. A fairly high ratio of chapter students to visiting students is necessary to keep everything running smoothly and safely. Our chapter students are charged with explaining the chemistry in terms that the visiting younger students can understand, and our students prepare handouts for them to take home to their parents for follow-up discussions. Our students soon discover that they understand the chemistry much better than they thought they did, especially when they successfully teach these concepts and are looked up to as role models. These events are often the most instantaneously rewarding; our chapter students immediately feel like the experts they are slowly becoming! The age difference also helps our students step into what feels like the professional adult world, since the students they serve perceive them as adults.

For older visitors we have been able to run full guided-inquiry style laboratories adapted from the *Journal of Chemical Education* and other laboratories that we run in our general and organic laboratories. These are especially exciting for our students since they are often running laboratories that are close to what they have just experienced and learned from in their own classes. Two significant differences in running these laboratories at different age levels are the amount of guidance students receive and the complexity of the process or scenario involved. For example, we run a number of CSI-style laboratories where students solve a mystery using chemistry (*9*). For middle school students, we would have only one or two pieces of evidence and one instrument involved; we might run a modified version of a drug components TLC laboratory from an organic laboratory manual with only four potential compounds and the additional use of GC-MS to confirm the identity of the unknown molecule (pure molecule only, no mixtures) (*10*). That same laboratory has been used with junior high school students with either mixtures of compounds to be separated and identified (with GC-MS confirmation) or with the use of only pure molecules and FT-IR along with GC-MS for confirmation of the identity of the unknown.

Visiting students have the experience of using real chemical instrumentation, which is exciting for them, and our chapter students have the experience of teaching students how to use the instruments and what the instruments can and cannot show us about the chemical structure of a molecule. Much of our laboratory curriculum is grounded in the use of instrumentation; chemists "see" molecules through spectroscopy, and these are the tools that chemists use to understand molecular behavior and quantify molecular composition. These laboratory experiences reinforce for our students the goal of "working like chemists" since they are using—and teaching the use of—the chemist's tools. No use of instrumentation and no reinforcement of concepts is too trivial; students learn something each and every time they acquire and analyze a spectrum, so the more exposure the better. The process of teaching these ideas and techniques is a fantastic reinforcement of classroom and laboratory concepts, "thinking like chemists", and has immediate results for our students. Students who help in these events inevitably perform better in our courses because they have a better understanding of the chemistry.

Taking the Chemistry to Them: Club Field Trips to Their School

While it is not as easy for our chapter to travel to another location as compared with a visit to our school, this is often the only way to reach classrooms that ask us for a chemistry experience. Some local grade schools and high schools simply do not have the ability to bus students to Utica College, while others cannot afford the classroom time lost to travel. With a large commuter population, there are many connections to our local grade schools and high schools so there is added incentive for our students to plan a visit. Depending on the needs of the classroom that invites us we plan either hands-on experiences or chemical demonstrations.

Chapter students are expected to help in the design of hands-on experiences that we take to local schools. Our goal of having students "think like chemists" by considering what they will need to have as they "work like chemists" is strongly reinforced in this process. Questions that they answer, with help from advisers, center again on safety and waste disposal. What do we need to bring to the school? Where will we be working and how can we do the experiment in that space safely? How do we dispose of the waste—what can be safely left at the school versus what we need to bring back to the college? We as advisers have a set of experiments that are well-tested and safe to take to schools, but we always have chapter students go through this process (even when we, as advisers, already know the answers) so that they have the experience of thinking through every step of the chemistry that they are taking outside our laboratories. Here again, students rapidly discover that they know—or can figure out on their own—more than they ever realized. This also helps them take ownership of the project and process, which makes them more enthusiastic and confident when it comes to the outreach day.

Some schools have been happy with whatever hands-on experience we can provide, which has allowed us freedom to choose whatever we think will work well and be interesting to the students. Chapter students have the freedom to pick the area of chemistry they love best and then find a way to make it useful and educational for the audience. Our students quickly feel like emerging experts in their sub-field after these events. For example, we ran a laboratory on extraction of DNA from onions, adapted from a laboratory text (*11*), for students at a local high school. There were approximately 30 high school students in a space that was a hybrid classroom/lab, and 10 of our chapter members planned and ran the event. Some of our students knew the high school science teacher, which made it easier for them to comfortably communicate with her, and were responsible for coordinating the supplies for the event. They made a full equipment and materials list, asking what was available at the school and determining what they needed to bring, and thinking of how to use the space safely and efficiently (for instance, chopping onions with household knives was appropriate at a desk, but all laboratory chemicals needed to be used only on laboratory benches). The knowledge that they needed to account for everything—if they did not bring an item, they would not have the item—led them to an increased appreciation for pre-laboratory assignments. Students who have participated in or run this event typically are much more organized and prepared in our advanced laboratories such as our quantitative analysis lab.

At other times schools have specific curriculum needs and ask us to provide experiences tied to those learning goals. These constraints are fantastic opportunities for chapter student learning. Instead of sticking with the old reliable experiments that are in their comfort zone, they are challenged to think about how chemistry really is the central science and can be applied to any science module an educator asks us to teach. For example, a local first grade class asked us to provide a hands-on experience for their light and energy module. We were not allowed to use flames, even in a demonstration, nor were we allowed to bring other heat sources (such as a hot plate) into the classroom. Chapter students were initially frustrated since they associated light and energy more with physics and all the "cool" chemistry experiments were either too advanced for first grade students or forbidden. With some guidance and brainstorming our students identified the chemical hook they wanted to give these topics: While we can't see individual molecules, we can see changes in chemical energy levels—what molecules are doing—through light (UV beads, glow sticks, UV-reactive ink).

Chemical demonstrations are always a chapter favorite, and our chapter is no exception. No matter how much first- or second-year students simply wish to run a "chemical magic show" it is standard procedure (and enforced by our officers, who have learned this wisdom) that all chemistry demonstrations need to be fully explained. It is chemistry, not magic, so the chemistry is central to anything we demonstrate. While every outreach we perform has students "communicating as chemists", chemical demonstrations really push chapter students into becoming effective communicators and chemical ambassadors. The general procedure is to plan 30–45 minutes of demonstrations in which one person runs the chemistry while another explains what is going on. Working in pairs increases safety and diminishes stage fright. We always run the same demonstration multiple times; we first show the chemical phenomenon, then explain what is going on chemically, and then show the demonstration again while pointing out the finer points of the physical or chemical change and reinforcing the chemistry involved. This approach works well with bigger groups and with any age range, and these events are particularly useful for retaining first- and second-year chapter students. Chapter members who spend one afternoon explaining and showing off their chemistry are very often hooked and keep coming back for more opportunities to demonstrate exciting experiments!

Reaching the Greater Community: Large-Scale Events

Large events, involving the greater community or the entire college community, are exciting and test the chapter's organizational and leadership abilities. We in our chapter define a large event to be one at which more than 100 people will be served and where the venue is open (outdoors, in a museum, or hall) and the audience is varied (entire campus community, attendees at the zoo, or other open venue). Students are required to brainstorm all the potential issues—especially safety considerations—that the venue and audience present: What if it rains? How many sets of goggles do we need? How do we explain this chemistry to a young child? To an adult who is scared of chemistry? To a nonscience-major friend? If we are preparing food, how do we prepare and keep the food safely and alert participants of potential allergens? All these variables seem daunting at first, but since we typically host at least one of these events a year there are students who have successfully organized these events and provide leadership and guidance to the chapter. For our students there is nothing like the satisfaction and pride that comes from bringing all the moving parts of a large-scale outreach successfully together.

One method of addressing the multiple moving parts issues has been for us to coordinate with an established large event and provide part of the experience for that day. The Central New York local section of ACS typically hosts large public outreach events to celebrate NCW. These events take place in the Museum of Science and Technology or the Rosemont Gifford Zoo, both in Syracuse, and are included with admission. Attendance at these events can run to well over 400 people. The local section takes leadership of some of the more challenging aspects (securing the space, local advertisement to the community) and our chapter, among others, provides the chemistry. Since these events require us to travel, students are again challenged to consider everything that is needed for the safe running of the experiments that day, reinforcing "thinking like a chemist" and "working like a chemist" and by the end of the event they truly have communicated like chemists.

Other events take place on the Utica College campus, which is an ideal way for chapter students to take pride in themselves and their majors. Hundreds of students and greater community members have attended Utica College's Taste of the Arts, a venue celebrating music, dance, visual, and written art. Our club celebrates the art of food and molecular gastronomy with fizzy marshmallows (chocolate-dipped homemade marshmallows sprinkled with pop rocks), brownies made with different variables modified (yielding cake-like or fudgy) and of course the liquid nitrogen ice cream. This table has always been highlighted in the local evening news as an example of creative and delicious science. One campus event that the club planned completely from start to finish was a NCW tie-dye event open to the entire campus community including 50 Young Scholars students. We provided 210 T-shirts with U-Ti-Ca in elemental symbols for coloring, as well as the dye for those shirts for those who brought their own (which represented another 250 participants), all other supplies, and even prizes. The process of organizing everything from fund-raising for the supplies, to stations for the tie-dying, to take-home instructions was an extremely useful one; after succeeding in such an event, even physical chemistry seemed easy to our majors!

Professional Development Events

While the events that we use to help nurture professional development are certainly not unique to our chapter, we believe the way that these events are promoted and integrated into our overall departmental learning goals to be a useful twist on these activities. We want our majors to become chemists; by meeting fellow chemists, listening to chemistry being presented and presenting in turn, or by simply accepting the role of chemical expert, our students discover a professional community and become part of it. The more we have been able to integrate these events throughout the entire major experience, the more likely we are to expose all our majors—not simply those active in our chapter—to the useful professional development activities our chapter has to offer.

Events at Which Students Are the Experts

In chemistry, sometimes it seems the more one learns the more there is to learn. Introductions to entire fields of chemistry (biochemistry, inorganic, physical) in junior or senior year show students just how wide and deep the chemical sciences are. This can be intimidating to students, leading them to undervalue the progress they have made since what they have learned is only a small percentage of what is out there to learn. Our mentoring activities center around the idea that mentoring is not only beneficial to those being mentored, it also reinforces to the mentor just how far they have come in their academic progress.

Our chapter holds formal events to help build these mentoring relationships, using the promise of both food and fun to draw in students. There are dedicated laboratory assistant safety training days at which junior and senior members bring the new laboratory assistants up to speed. Typically these days end with pizza, an ice cream run, or a group outing to a movie. During reading days, we have had a study room called "Will Think for Food" where science faculty are encouraged to stop by and get snacks and answer questions, and chemistry tutors help out general chemistry and organic students as well as nonmajors for our introductory courses. We have a constant pot of coffee going, the college buys doughnuts for us, we buy pizza, and everyone donates some baked goods. During the day there are interactive events such as the much-loved Jeopardy review session for general chemistry. By the end of these events our first- and second-year members, as well as other students in our college community, have learned a great deal about chemistry. Our junior and senior chapter members have realized just how much they have learned; they were once the students struggling in organic, and now they are the students who understand it well enough to explain it.

Working as judges for either our own Utica College Regional Science and Engineering Fair or ones hosted by our local ACS section has been a great experience for our students. This is a place where all our chapter members, even those in their first year, can work as experts. We remind them before the event that a sixth grade student will see little difference between someone just starting their college career and a senior; to them, we are all chemical experts coming to talk about their research project. What the sixth grader will remember is how engaged we were as an audience, how excited we were at their results, how encouraging we were during the question and answer portion. All our students have the minimum knowledge necessary to serve as judges and as good audience members. What matters most is within their ability to control; being an engaged, excited, and encouraging professional. When our members are seen as professionals by the students, when it is clear that their opinions matter to the students presenting their projects, when they are trusted with assessing the success of the project and

13

presentation, they begin to realize that they are part of a professional community and are young professionals themselves.

Meeting the Chemical Community

Students continue to be amazed when they are told that, by joining the American Chemical Society, they are joining the largest scientific organization in the world (*12*). While chemistry might seem intimidating as a major to freshmen, the simple knowledge that they are not alone—that there are a lot of successful chemists out there doing amazing things—opens up so many possible paths. Taking students to professional events is not something reserved for the best students in the course, but instead is integrated into the chapter and the departmental missions; we want our students to become chemical professionals, and so we take them to meet all the chemical professionals that we can. Since our discipline and our community are so large, we can—as educators and as chapter advisers—find a place for every student to fit in. This helps our chapter, our department, and all our majors.

Our student chapter became reactivated through a trip to NERM in Binghamton (2006) where students essentially demanded "more of this" while pointing and waving at the meeting; they wanted not only the speakers and the chemistry, but the chemical community they saw in action. They saw chemists writing equations they could follow on placements, they heard geeky jokes they understood, and they were treated with respect and their questions and ideas were listened to by chemists and biochemists they admired. They peeked into a world that they decided they loved, and they wanted in.

We have worked hard as a chapter and a department to keep that sense of wonder alive. Our chemical community is a network rich with opportunity and support, and is a resource beyond any other. All that is needed to be part of it is to become part of it. We of course attend the larger events as a chapter, such as NERM and National Meetings, but we also use this network to help arrange smaller events such as visiting speakers, local section meetings, or tours such as brewery tours or tours of local industry. Planning these smaller events helps our chapter officers see the process come full circle. As seniors, they are planning events that help encourage first- and second-year students to see the possibilities that our science opens up for them. The seniors also see the chemical professionals, some of them program alumni, come back to show the seniors what they have achieved. Chemistry helps open doors for them that they did not even know existed before they started college, and once they are through those doors they have successfully joined the chemical community.

References

1.   Utica College Website. http://www.utica.edu/instadvance/marketingcomm/about/ (accessed June 19, 2018).

2. van der Zanden, P. J. A. C.; Denessen, E.; Cillessen, A. H. N.; Meijer, P. C. Domains and predictors of first-year student success: A systematic review. *Educ. Res. Rev.* **2018**, *23*, 57–77.

3. Thomas, A. C.; Boucher, M. A.; Pulliam, C. R. Qualitative to Quantitative and Spectrum to Report: An Instrument-Focused Research Methods Course for First-Year Students. *J. Chem. Educ.* **2015**, *92*, 439–443.

4. Pulliam, C. R.; Pfeiffer, W. F.; Thomas, A. C. Introducing NMR to a General Chemistry Audience: A Structural-Based Intrumental Laboratory Relating Lewis Structures, Molecular Models, and ^{13}C NMR Data. *J. Chem. Educ.* **2015**, *92*, 1378–1380.

5. *inChemistry*, November/December 2008, Vol. *18*, https://www.acs.org/content/dam/acsorg/education/students/college/inchemistry/in-chemistry-december-2008.pdf (accessed June 19, 2018).

6. Boucher, M. Fighting Self-Doubt When Imposter Syndrome Kicks In. *inChemistry*, Feb 20, 2017, https://inchemistry.acs.org/content/inchemistry/en/college-life/imposter-syndrome.html (accessed June 19, 2018).

7. ACS Educational Resources webpage. https://www.acs.org/content/acs/en/education/outreach/ccew/educational-resources.html (accessed June 19, 2018).

8. Shakhashiri, B. Z. *Chemical Demonstrations: A Handbook for Teachers of Chemistry*; University of Wisconsin Press: 1983–2011; Volumes 1–5.

9. Specht, K. M.; Boucher, M. A. A Froensic-Themed Case Study for the Organic Lab. *J. Chem. Educ.* **2009**, *86*, 847–848.

10. Lehman, J. W. *Multiscale Operational Organic Chemistry*, 2nd ed.; Prentice Hall: New Jersey, 2009; 147–151.

11. Bettelheim, F. A.; Landesberg, J. M. *Laboratory Experiments for Introduction to General, Organic and Biochemistry*, 8th ed.; Cengage Learning: 2016; pp 515–518.

12. ACS Fast Facts webpage. https://www.acs.org/content/acs/en/about/aboutacs.html (accessed June 19, 2018).

Chapter 2

Staying the Course

M. T. Blankenbuehler* and B. Van Ness

Department of Biology and Chemistry, Morehead State University,
150 University Blvd., Morehead, Kentucky 40351, United States
*E-mail: m.blanken@moreheadstate.edu

The student chapter of the American Chemical Society (ACS) at Morehead State University (MSU) has been recognized with awards from the ACS almost every year since 1995. The focus of the club changes from year to year depending on the areas the officers would like to emphasize. While tradition is essential to the long-term health of the club, variation and creativity are highly valued. The organization has been involved with a myriad of activities in the areas of outreach, service, fundraising, professional development, philanthropy, and fun (i.e., social activities). Activities over the years have included giving the president of the university a pie in the face to raise money and awareness for the American Cancer Society, cleaning local hiking areas, tutoring students in chemistry, conducting demonstrations for area elementary, middle, and high school students, holding raffles to raise money for national meeting trips, hosting speakers and alumni to talk about career paths, running 5K races to support Alzheimer's research, taking trips to area industries and seminars, and, of course, selling periodic table cupcakes during National Chemistry Week. This chapter, like all others, has its challenges but has managed to "stay the course" by promoting chemistry and doing its best to excite, inform, and engage those individuals interested in science.

Introduction

Morehead State University (MSU) is a public, regional university with an enrollment of about 10,000 students. It is located in the Daniel Boone National

Forest in rural eastern Kentucky. Morehead Normal School was originally founded in 1887 and became Morehead State University in 1966. The chemistry program is part of the Department of Biology and Chemistry and has approximately 60 students, with an average graduating class of four to seven majors each year. The vast majority of the chemistry majors are pursuing preprofessional studies for careers in health fields such as medicine, pharmacy, optometry, and dentistry. The chemistry program's faculty consists of five tenured/tenure-track faculty, one instructor, and one staff member (lab coordinator/instructor).

The earliest record of the Student Affiliates of the American Chemical Society (ACS) dates to 1969 as a Mu Sigma Chi chapter. This chapter required the completion of general chemistry. Another chemistry club, Beta Chi Gamma, which included math and physics, has also been referenced (1). Dr. Mike Adams, currently at Xavier of Louisiana, reenergized the chapter in the early 1990s and was able to put the chapter back on the path to success. He left Morehead State in 1998, and Dr. Zexia Barnes took over during 1998 and 1999. Dr. Mark Blankenbuehler took over in 2000 and is still chapter advisor. Dr. Brandon Van Ness has recently come on board as a co-advisor. The chapter has seen regular success since 1995.

ACS recognition since 1995 includes the Commendable Award in 1995, 2003, 2006, 2007, 2009, 2010, 2011, 2012, 2013, 2014, and 2015 and the Honorable Mention Award in 1996, 1997, 2001, 2004, 2005, 2008, and 2016. The Certificate of Achievement was awarded in 2002 and the Green Chemistry Award in 2013.

Success over Time

What Makes It Possible for a Student Chapter To Stay Active and Maintain the Effective and Appropriate Level of Activity?

Of course, every chapter is different, and various strategies may be employed to keep a chapter on track and moving forward. The following should be considered when deciding how best to help put the chapter on the right path.

Who Should Serve as the Faculty Advisor?

Consistent, energetic, and effective leadership is essential. Finding the correct person to serve as the faculty advisor makes all the difference.

The advisor must be familiar with university regulations regarding chapter functions, which includes rules for everything from fundraising, travel, and the safety of students and the public to finances, communication, and food usage. A student organization handbook should be available, but the first resource is always the faculty advisor. The advisor can provide the best guidance to help students determine what can and cannot be done and how best to accomplish their desired goals according to the regulations set forth by the university. Some organizations may try to operate under the saying "It's better to ask forgiveness than permission," but this is a slippery slope that should be avoided.

Patience is a virtue and the advisor must be well-versed in it. Chapter officers are still students, and as such, are still learning how to become leaders, and most do not possess these skills before taking on the mantle of peer leadership. Supportive coaching and instruction are necessary tools that the advisor must employ to develop each student officer as an independent, confident leader.

Just like any student in any class, there are times in which the student officers will let the chapter or advisor down. While these events are rare, there are two schools of thought for the faculty advisor to employ. The advisor can pick something up and get it done, or let the officers fall flat on their faces to prove a point. An advisor is supposed to be a mentor to the students and develop their leadership skills and foster an environment focused on responsibility. The old adage "Leaders lead from the front" is something that should be remembered by the faculty advisor.

Many variations exist on the way to produce effective chapter leadership, and the advisor must be willing and able to adapt. The advisors are not there to run the clubs and do everything themselves, although every now and again, it will feel that way. Guide the students and assist them whenever they need it. Prod them when needed and be ready with a swift kick in the pants for failures, and everything will work out for the best.

The advisor must have a real interest in seeing the chapter succeed and the students in the chapter grow and learn. Taking on this service activity requires long-term dedication. Energy and enthusiasm are required to get the most out of the students and pull off the kinds of activities that impact the department, campus, and community in positive ways. Often, great advisors were once ACS student chapter officers as well. Dr. Blankenbuehler served as the vice president and president of his undergraduate institution student chapter. The opportunity to serve in that capacity had a lasting impact on him. He learned many valuable lessons, but perhaps most important was how an engaged, enthusiastic, knowledgeable faculty advisor (Dr. Sally Hunnicutt) can make all the difference in a successful student chapter of the ACS.

Who Should Serve as Officers? What Should Be the Method of Elections?

This, of course, varies depending on a variety of factors. The way this has been conducted at MSU has not varied much over the years.

A formal election conducted with the ACS members has occurred only once since 2000. The MSU chapter has been able to fill all officer positions through a more informal process. Students interested in becoming officers discuss this with the current officers and the advisor. Only students who have been active members the previous year and student members of the ACS are eligible for a leadership position. The student must be involved in a majority of the club events to be considered active. Formal documents detailing the duties of each position in the chapter are available, and they are shared with each prospective officer candidate. Those students who are qualified and desire to serve as officers then gather in committee to discuss in which positions they want to serve. If, after the initial discussions, more than one person wishes to run for the same office, elections are held. The norm is that the students agree to serve in different positions and all that is required is approval by acclamation. This can be an entirely new set of officers each year, which can cause some discontinuity. Every effort is made to encourage active freshman to serve as officers when they are sophomores, but usually, active sophomores serve when they are juniors. Having an entire slate of officers who are seniors may produce a very active year, but the next year is a huge challenge. There are no automatic officer tracks, such as vice president serving as president the next year, although the vice presidents typically become presidents the following year.

The chapter has experimented over the years with additional officer positions with varying amounts of success. Typically, the chapter has the four standard positions (president, vice president, secretary, and treasurer). Additional experimental officer positions have included public relations, historian, fundraising, community service, and co-vice presidents. These attempts were made to provide additional weight to the position to encourage those in these experimental positions to be more active. This initiative was ineffective and was abandoned. The chapter usually leaves the leadership of the areas that fall outside the typical duties of the four standard officer positions to that of committees, in which a student can act as chair.

How Can the Chapter Become Financially Stable and Maximize Its Impact for Chapter Members, the University, and the Local Community?

While obtaining random grants or having fundraisers for specific individual activities, events, or travel is admirable, the development of consistent funding sources, whether ongoing grants or dependable, regular fundraisers, is essential to the financial survival of the organization.

The primary fundraising for the MSU chapter is through the sales of chemistry laboratory manuals and laboratory aprons and goggles at the beginning of each semester. With approximately 600 total students enrolled in general and organic chemistry courses, this fundraiser serves as a solid foundation of regular funding. At the end of each semester, unwanted and used aprons and goggles are collected in each laboratory section from those who wish to donate them to the chapter. These used aprons and goggles are recycled and sold at a discount the next semester.

Other regular fundraising efforts include the selling of ACS exam study guides (2) and obtaining ACS travel grants (3). Typically, the chapter can sell about 60–75 ACS study guides each semester, which makes this a modest fundraiser but still regular income. Every year, the chapter applies for a $300 travel grant from ACS for students to attend the spring national ACS meeting. Other fundraisers are more random and are held at the discretion of the officers and membership. Recently, a new fundraiser was attempted and determined to be beneficial for the chapter, which will lead to its use more often in the future. Local restaurants allow the chapter to collect a portion of the proceeds on a given day. This usually involves minimal work on the part of the chapter by contacting the restaurant and setting up the event with management. These have typically raised about $150 per occurrence. Fundraisers such as raffles; bake sales; and t-shirts, jogger pants, and graduation stoles sales have been conducted with varying success and generally raise from $50–150 each.

Occasionally, funds from the Department of Biology and Chemistry, the Dean of the College of Science, or the Student Government Association are also obtained. These funds are specifically for travel for students presenting research.

Who Should Make Up the Membership of the Chapter?

Maintaining a critical mass of engaged members who are not officers is essential. No local dues are collected and chapter membership is not restricted by major, although most of the members are chemistry and biology/biomedical science majors. All students are welcome. Students are, of course, strongly encouraged to join ACS and are given pamphlets or information sheets that describe the ACS and its benefits at general meetings. Each year, the chapter typically raffles off four to six student memberships to ACS to improve the retention of student members. Typically, the MSU chapter of the ACS has 15–18 paid ACS members.

Over the years, various efforts and strategies have been employed that focus on improving the recruitment of new freshman members. The chapter participates in university organization fairs during orientation or the first week of classes in the fall. Chapter members visit the general chemistry courses early each semester to put the chapter on new students' radar. Regular chapter meetings are held every two weeks and have had a positive impact on overall member retention. The chapter meetings in which free food is provided are always the best-attended meetings.

Specific efforts that encourage students to become members from the earliest possible opportunity lead to a more stable membership composition and enhance leadership transitions. Although these efforts have not yielded an increase in membership over the years, they have been effective in keeping the membership stable. It has been observed that membership tends to dip in years when some of these activities are not done.

Year-to-Year Activity

Each year brings a new group of officers and a new set of ideas for activities. The officers and members are free to engage in activities similar to those conducted in the past but are not required to do so. Students are more likely to "buy in" and be enthusiastic about activities they have an interest in rather than just being told to do the same activities every year. Many of the activities from the previous year are scheduled each year because the students enjoy them, but the point is made that the yearly activities are not scheduled for the members in advance.

Service to the Community, University, Local Section, and Department

Many activities focus on engaging the community at all age levels and occur at varying times throughout the year, including during National Chemistry Week (4).

For many years, efforts were made to conduct demonstrations or laboratory activities in local K–12 schools throughout eastern Kentucky. The focus of the activity changed depending on the grade level. The term "magic show" was used for younger students, whereas more-detailed scientific activities were mixed in when recruiting at the high school level. In more recent years, greater limitations have been imposed by local school districts on the materials that the chapter members could bring into the schools. Although there are many demonstrations and activities that use only household chemicals and do not make any loud noises or create interesting odors, the focus had been on exposing the students to things they had not experienced or would have difficulty obtaining the supplies to do themselves. Recently, the chapter members have focused on bringing the K–12 students to MSU to overcome these obstacles in the local school districts. This shift enables chapter members and faculty to conduct demonstration activities without the stringent limitations on chemicals and equipment set forth in the local school districts and allows the chapter members a longer time period for engagement with these students. Elementary students often spend the day at MSU and conduct activities in physics, math, biology, and chemistry, resulting in a much more immersive experience than a single visit to their K–12 classrooms by chapter members. Students in the Rowan County Gifted Program were able to perform a limonene extraction from orange peels using supercritical CO_2 at MSU—an experiment that would have been impossible to perform at their school. High school students can be engaged in more-advanced experiments, such as using gas chromatography to separate a mixture of alcohols and infrared spectrometry to determine the differences between the functional groups of cyclohexane, cyclohexanol, and cyclohexanone. Informative sessions on careers in chemistry also can be conducted.

Recognizing the needs of current MSU students has been a major focus of the chapter. Advanced students who have successfully completed chemistry courses are recruited each year to be tutors. Students apply to be tutors and a faculty member advises the tutor on some things to do and not to do during tutoring. Many students have extreme difficulty in chemistry courses and the chapter tutors provide a valuable service by tutoring those students for free. Typically, chapter members provide more than 400 hours of free tutoring per semester. Tutors are provided for all introductory chemistry courses and organic chemistry including survey courses for nonmajors. Occasionally tutors for advanced physical, inorganic, and analytical chemistry can be provided.

The chapter has also found ways to financially support the university and community. The regularly conducted fundraising permits the chapter to fund the Richard Hunt Memorial Scholarship and purchase chemistry equipment as needed. The scholarship, solely funded by the chapter, provides up to $400 per semester for a qualified chemistry major to use on textbooks. Donations are also made to a variety of local and national organizations depending on member preference and fundraising success. Recent cash donations to local organizations have included the Rowan County Soup Kitchen (5), The People's Clinic (6), and the Blessing Hands organization (7). Blessing Hands supports educational outreach efforts in China. The chapter was able to sponsor a Chinese student interested in science and also provided age appropriate science books to a Chinese elementary school library. National organizations, such as the American Cancer Society and the Alzheimer's Association, have also been supported.

Because MSU is located in the Daniel Boone National Forest, the chapter regularly engages in efforts to keep the area clean. Each year, the chapter members perform organized clean-up activities that remove debris from popular hiking areas, which creates greater awareness for environmental stewardship. These activities are often conducted during Earth Week (8). The activities provide opportunities to collaborate with other organizations at MSU. Most of the club members are also members of other related clubs in biology, physics, and geology. Those members always seek out opportunities to work together on appropriate projects.

What Is a Chapter without Some Fun?

Activities reserved for members only are important to building community in the chapter and enhancing retention and recruitment efforts. Each semester, picnics are held at either a local shelter or at Cave Run Lake. These picnics provide a chance to eat, talk about future plans, or take in a game of volleyball, basketball, bocce ball, or frisbee golf. Potluck events, movie nights, elemental bingo, white elephant gift exchanges, intramural volleyball, soccer, and basketball, or forming a team to run in a local 5K race are all great ways to have fun.

Professional Development

Efforts to enlighten chapter members regarding career opportunities and also build skills that will be beneficial beyond the student's time in higher education are important.

One event that chapter members really enjoy is alumni night. MSU graduates are invited back to talk about their experiences in their professions. Individual and panel discussions are typical. Most of the returning alumni are in professional schools and provide valuable insight into the application, interview, and acceptance process. A variety of alumni are always invited to ensure students have someone in their career path to speak with.

Chapter members also invite local scientists to present their research. This provides an opportunity for members to invite speakers with areas of research of interest to them. Opportunities are also made for students to talk with speakers about their research, graduate opportunities, and career opportunities.

Efforts are made to attend ACS Lexington local section meetings (9) as well as the national ACS meeting to provide opportunities to learn more about careers in chemistry and to network. However, participation in local section activities are rare due to the 75-minute drive to Lexington and the fact that those events are generally held during the week at 6 p.m. Faculty and students are not done with labs until after 5 p.m. most days. Still, the occasional carpool trip is made.

Students who conduct research at MSU present a research poster as well as a successful chapter poster at the national ACS meeting every spring. Depending on interest and financing, chapter officers and other chapter members make the trip as well. Financing the trip, which can be a substantial investment, is a challenge. Hotel costs are covered by the regular fundraising efforts. Students going on the trip conduct additional focused fundraisers. Departmental support for registration fees is also typical for those presenting research. Additional funds are occasionally obtained from other campus sources, like the Office of Research and Sponsored Programs and the Student Government Association. Getting to the meeting, however, is usually the responsibility of the student. Almost every meeting requires a flight from Kentucky. Every effort to defray the costs for students is made to enable them to experience this important event. Shorter trips to local industries for tours are also popular and give members a first-hand look at the facilities and the opportunity to learn more about future career paths.

Conclusion

The ACS chapter at MSU has managed to create a culture of success over the past 20-plus years. The road has never been easy, but it has been rewarding for the faculty and students. While faculty advisors are an important resource, ultimately, the students have made it happen. Their work ethic and enthusiasm for science are to be commended.

References

1. MSU Raconteur 1969 – Morehead State University Online Archives Web Site. https://scholarworks.moreheadstate.edu/morehead_state_yearbooks/ (accessed April 12, 2018).
2. ACS Exams Institute Web Site. https://uwm.edu/acs-exams/ (accessed April 12, 2018).
3. ACS Travel Grants Web Page. https://www.acs.org/content/acs/en/ funding-and-awards/grants/acscommunity/studentaffiliatechaptergrants/ travelgrants.html (accessed April 12, 2018).
4. ACS National Chemistry Week Web Page. https://www.acs.org/content/acs/ en/education/outreach/ncw.html (accessed April 12, 2018).
5. Community Soup Kitchen Facebook Page. https://www.facebook.com/ CommunitySoupKitchen/ (accessed April 12, 2018).
6. People's Clinic Web Page. http://www.peoplesclinicmorehead.com/ (accessed April 12, 2018).
7. Blessing Hands Web Page. https://sites.google.com/a/blessing-hands.org/ blessing-hands/ (accessed April 12, 2018).
8. ACS Earth Week Web Page. https://www.acs.org/content/acs/en/education/ outreach/ccew.html (accessed April 12, 2018).

9. ACS Lexington Local Section Web Page. http://acslexington.sites.acs.org/ (accessed April 12, 2018).

Chapter 3

Forming Bonds: Using Community Outreach To Maintain Relationships with Other Chemistry Societies

Leanna C. Carter, Kearney M. Foss, Debra L. Mohler, David L. Wilson, and Isaiah Sumner[*]

Department of Chemistry and Biochemistry, James Madison University, 901 Carrier Drive, MSC 4501, Harrisonburg, Virginia 22807, United States
[*]E-mail: sumneric@jmu.edu

The Student Affiliates of the American Chemical Society (SAACS) Chapter at James Madison University (JMU) is a strong advocate for chemistry both on campus and in the surrounding communities. The main way that SAACS fulfills this mission is through chemistry demonstration functions. In addition to hosting several small shows each year, SAACS leads two major community-based demos: Skeleton Festival in the fall semester and ChemFest in the spring semester. In this chapter, we describe the planning process and the improvements for the main events that we conduct throughout the year, the importance and role of a department outreach coordinator (DOC), and how community outreach helps our SAACS chapter maintain a fruitful relationship with other student groups in our department.

Introduction

James Madison University (JMU) is a comprehensive, coeducational, primarily undergraduate institution located in the Shenandoah Valley of Virginia. Founded in 1908 as The State Normal and Industrial School for Women at Harrisonburg, JMU is now a highly competitive, state-supported institution with an enrollment of nearly 21,000 undergraduate students and 1900 graduate students primarily in professional programs and 925 full-time faculty. The university is committed to the inseparability of scholarship, research, and teaching.

The Department of Chemistry and Biochemistry has 27 full-time faculty members, 21 of whom are research-active. All faculty in the department are dedicated to providing the best education possible to our students, and we recognize the great importance of teaching through research. The department is certified by the American Chemical Society (ACS) and offers Bachelor of Science degrees with ACS certification in chemistry with concentrations in chemistry, biochemistry, material science, and chemical education. Additional degrees are offered with general and business tracks. The department began to offer a new biophysical chemistry major in the fall semester of 2013. There are no graduate students in the department, nor are there any plans for a graduate program. The department currently graduates roughly 30–40 majors per year, with about 200 chemistry majors in total.

There has been a thriving chapter of Student Affiliates of the American Chemical Society (SAACS) at JMU since the mid-1970s. SAACS (1) was founded to bring undergraduate chemists together, socially and professionally. SAACS social events have included organization picnics, intramural games, and holiday parties. SAACS professional events have focused primarily on touring companies to see how chemistry can be applied in the industrial and medical fields. As of 2018, SAACS at JMU has 40 members and an executive board consisting of a president, a vice-president, a secretary, a treasurer, an outreach coordinator, a tutoring coordinator, and a social events coordinator.

Alpha Chi Sigma (AXΣ) (2) is a chemistry fraternity that was started at the University of Wisconsin at Madison on December 11, 1902, and strives to advance chemistry professionally and socially. It has been coeducational since 1970. The first chemistry fraternity at JMU was formed in 1989 by two chemistry students who sought to be part of Greek life with a chemistry focus, and they named the fraternity Kappa Sigma Mu (KΣM, pronounced "chem"). KΣM became the foundation for AXΣ to launch a chapter at JMU. The Gamma Kappa chapter of AXΣ initiated its first class on April 20, 1991, and has been inducting new members every year thereafter. The current chapter of AXΣ has 60 members and officers including a president, a vice-president, a master of ceremonies, a treasurer, a reporter, a recorder, a historian, an alumni secretary, a webmaster, and a standards chair.

Although both organizations are chemistry clubs, they have different emphases and benefits. SAACS, like the ACS, is primarily a professional organization. Students can join at any time, starting with their first semester, and are not required to become members of the national ACS. Members benefit from the professional networking and development sponsored by SAACS and by the association with the larger ACS network. AXΣ is primarily a social organization. Students must be in their second year before joining. However, membership is lifelong and even faculty members can join and participate. Members benefit by being able to network with other fraternity "brothers" at social events. Many students are both SAACS and AXΣ members.

Despite the differences between the clubs, there is overlap between their missions. Furthermore, the prospective membership for each club is the same, that is, JMU chemistry students. As a result, shortly after AXΣ's founding at JMU, SAACS and AXΣ became competitive and developed an antagonistic relationship, with disputes primarily over membership and fund-raising activities. Because the department recognized the benefit of both clubs and because this conflict was having an adverse effect on both organizations and on the department, the leadership of the two groups drafted a resolution to maintain "professional and collegial relations" and to "cooperate to enhance all aspects of chemical education at James Madison University". This document, known as the Miller Accords (named after the building housing the department at the time), outlines the responsibilities of each organization to the student body, ensures that fund-raising efforts are not duplicated, and provides a mechanism to resolve disputes. The Miller Accords are considered a living document and are reaffirmed annually by the AXΣ and SAACS leadership. In the original accords, it was stipulated that students must be members of both organizations, but that they could serve in the leadership of only one at a time. The first provision prevented one club from monopolizing prospective membership and the second prevented conflicts of interest. More recently, the Miller Accords were amended, so that AXΣ handles lab notebook sales and SAACS is responsible for organizing tutoring and selling lab goggles and study-guides. These provisions make sure each group is able to fund-raise effectively. (A current copy of the accords can be found in reference (*3*).)

The Miller Accords have provided a formal outline for how AXΣ and SAACS should cooperate. However, we have found that coordinating and running chemistry demonstrations is an excellent, informal way to maintain amicable relations between the two organizations. The students in both clubs are passionate about chemistry and love to advocate for chemistry and science in the community. In fact, each club requires that members participate in at least one demo per semester to remain in good standing. Therefore, each group also has its own student demonstration coordinator. However, these coordinators work with each other and with the department's independent faculty coordinator, Dr. Wilson, to plan and execute joint events. By working closely with AXΣ on annual community demos, we have found that the groups see each other as valuable partners. Herein, we describe several community shows that are organized by SAACS with AXΣ participation, and we highlight how the demos have benefitted from the cooperation between the organizations. We also describe the creation and role of an independent (i.e., not affiliated with AXΣ or SAACS) department outreach coordinator (DOC). By ensuring that such events are coordinated independently, demos are no longer seen as belonging to either organization, but to the department.

Department Outreach Coordinator

The Department of Chemistry and Biochemistry at JMU coordinates, oversees, and participates in a number of different types of outreach events: high school visits to the department, faculty outreach to individual classrooms or whole schools, and SAACS/AXΣ-run events, for example. It is the responsibility of the DOC to provide support for these events, to act as a department contact for off-campus entities that request events, and to monitor the type and quantity of events involving the department. This includes SAACS and AXΣ demo shows. With no affiliation to either organization, the DOC has the flexibility to assist any person or group in the department involved in outreach and to help integrate all outreach efforts.

There are several advantages that an independent DOC provides to SAACS (and AXΣ). One is continuity. Since the coordinator is a permanent faculty member, he/she can help us establish long-term goals for our outreach efforts and can help us pursue those goals over a period of years, even as member cohorts graduate. The DOC can also marry those long-term goals to the department's long-terms outreach goals. In this same vein, the coordinator helps us retain institutional memory as different cohorts take over for one another. As anyone who advises student groups knows, not all cohorts are equally engaged and active. An independent coordinator can bridge the divide between cohorts as groups with less interest in outreach move through the department. This will prevent new, more engaged groups from having to relearn lessons that were already mastered by previous classes. In addition, the DOC can motivate less interested classes as they matriculate through the department. To prevent a slump in outreach from the department, with a handful of students the coordinator can carry on outreach efforts on behalf of the department until a sufficiently active cohort of chemistry majors arrives. In addition, the DOC is in a position from which he or she can maintain and foster the outreach networking contacts that we have developed. Thus, schools and local organizations have a stable, permanent contact in the department with whom they can communicate.

Another advantage granted by this position is the perspective that it can bring to our organization. For example, prior to the creation of the DOC, there were no data being collected describing the extent of the department's outreach efforts. Now, the coordinator collects geographic and demographic information, which allows the department to compile a yearly report on the extent of their outreach efforts. This helps us and the department identify areas and populations into which we can expand our outreach. In addition, with the department informed of the extent of outreach, the coordinator can advocate for more department support and can apply for external funding to support outreach. The coordinator can also use his or her experience in performing demos to help us refresh stale demos or to find new demos to replace old ones. Shortly after this position was created, the SAACS and AXΣ demonstration organizers and participants had become bored because they had given the same performance for years. Therefore, the DOC organized a literature search at the campus library with interested SAACS and AXΣ members, which gave them a number of new ideas for shows. Due to the success of this effort, they now explore new demos for each event.

In addition to seeking out new demonstrations, the coordinator is also instrumental in incorporating a more coherent set of learning objectives into our performances. At present, there is little peer-reviewed research on the effect of such shows on learning chemistry, so we began to develop learning objectives for them. These included improving the communication between volunteers and the audience about the science involved in the demos, developing training videos for new volunteers to show them how to perform and to explain the observed phenomena, and making an effort to tailor the explanation of the relevant chemical concepts to the age/experience of the audience. With the training videos, demo instructors are able to evaluate their volunteers' mastery more carefully. Previously, trainees were required to watch a demonstration being performed, then immediately repeat what they had seen. This method did not lead to long-term retention of what they had learned. More recently, we have changed the process and we ask the trainees to study the videos first, then to execute what they learned in front of experienced demo performers. In this way, we hope that the trainees will transfer their understanding to long-term memory more efficiently. We are still evaluating this method.

We take pride in the fact that our student organizations are run entirely by the student body; all SAACS and AXΣ events are organized, managed, and produced by students. (The department provides storage space for chemicals we purchase, oversight to ensure safe storage of chemicals, and oversight to ensure safety precautions are taken during demonstrations.) However, faculty participation is still crucial to running a successful organization. For example, faculty advisors have had to make timely interjections, as they did when the Miller Accords were drafted. Additionally, we have come to see the advantages of an independent faculty advisor. The DOC can provide continuity and perspective for both SAACS and AXΣ without being seen as biased toward either group. The coordinator is available to both groups and any other member of the department involved in outreach. This position helps to coordinate, facilitate, and integrate SAACS's outreach goals set with those of AXΣ and the department. Finally, the DOC helps maintain the neutral status of demos and helps maintain the symbiotic relationship SAACS has with AXΣ.

Community Demonstrations

The schedule of demonstrations in a typical year includes two major events (i.e., ChemFest and Skeleton Festival) and many smaller ones. Since ChemFest and Skeleton Festival are quite large, they require the greatest commitment of time and resources from SAACS, AXΣ, and the DOC. Therefore, these functions are purposefully scheduled in different semesters: spring and fall, respectively. In contrast, smaller shows are scheduled throughout the year as negotiated with partnering organizations (i.e., teachers, schools, day cares, other university departments, etc.).

We start planning for ChemFest and Skeleton Festival three to four months in advance. A committee is formed and it schedules weekly preparation meetings. In the first month, the committee decides which demos to perform and then the committee members break off into subgroups to learn them. The committee spends the next two months training volunteers to learn the procedures. The last couple of weeks are spent gathering and purchasing supplies, and the week before the event all of the supplies are sorted into tubs according to the demonstration so that each station has all of the required materials. The tubs are put in place an hour before the shows start.

ChemFest

ChemFest is a large event that attracts hundreds of school-age children from the local communities. It combines multiple student-centered, hands-on activities divided among different stations with big presenter-centered demo shows. The aim of ChemFest is to get students excited about chemistry. This event differs from the typical productions done by JMU SAACS because of the size of both the production and the audience.

AXΣ started ChemFest in 2011 when Westwood Hills Elementary School in Waynesboro, VA (about 30 miles away from campus) asked the organization to hold a demo day at their school. ChemFest has since become an annual event. (ChemFest 2014 was canceled because of scheduling conflicts.) In the spring of 2016, AXΣ was unable to host ChemFest due to a change in leadership and in the funding available for the event. Because AXΣ recognized the importance of ChemFest to the community, they wanted to ensure that it continued. Therefore, AXΣ asked if SAACS would be able to step in. This transition was possible because of our good rapport.

ChemFest 2016 and 2017 were held at Westwood Hills Elementary School (ES) for the students and their families. The event followed an annual charity 5K run that Westwood Hills ES hosts. The overlap of these events boosted attendance since the 5K participants and their audience could simply stay and enjoy the demos. ChemFest 2018 was relocated to James Madison University to increase the variety of demonstrations performed without raising safety risks and to allow us to host a larger audience. We also expanded the audience beyond the Waynesboro community and invited other local elementary, middle, high school, and college students. Since ChemFest 2018 was on campus, we had the ability to perform demos appropriate for all grade levels.

Although SAACS has coordinated ChemFest since 2016, we continue to work closely with AXΣ and the Chemistry Department. ChemFest started as a two-hour event, but it continues to lengthen every year. Therefore, it is crucial for SAACS to work with other student groups to be able to effectively execute this event. Working with other organizations and the department also makes it easier to fulfill the number of volunteers needed to run the event. In 2016, there were about 30 students to run the event; however, in 2018 there were 40. The number of volunteers grows as the event gets bigger because more hands are needed to help it run smoothly. The volunteers help set up, run the stations, perform the stage shows, and help with cleanup. Roughly ten people are needed to set up, two people are needed at each station, one person is needed to perform each demonstration during the stage shows, and all volunteers present are needed for cleanup. ChemFest counts as a required semesterly demo for students in both SAACS and AXΣ, and it counts as seminar credit for first-year chemistry majors. This event is important for the first years because it familiarizes them with the student organizations in the department and provides a mechanism for them to get involved.

Two types of demonstrations are performed at ChemFest: interactive, hands-on demos that are set up on tables and performed continuously, and two stage shows that showcase more advanced chemistry where safety is a bigger concern. We typically perform four to six experiments during the stage shows. They start 30 minutes after the event opens and occur periodically until 30 minutes before close. Some of the hands-on demonstrations we have used are shaving cream art, gummy worms, silly putty, balloon skewers, dry ice bubbles, and candy milk rainbows (*4*). The stage show demonstrations include a "Magic Eraser" flame test (*5*), a rainbow tube, a liquid nitrogen cloud and breaking fruits, and glowing pickles (*4*). Because ChemFest 2018 was held at JMU, we were able to have the proper safety equipment (e.g., fume hoods) for the higher level demos, such as methane bubbles, gummy bear sacrifice, and extinguishing ear wax flames (*4*).

SAACS constantly strives to improve the safety, efficiency, and impact of ChemFest. For example, we have purchased more goggles for the students to borrow during the demos, because attendance always seems to exceed expectations. We also learned that it is important to have a surplus of supplies for the demos that require consumables; one year we had to close some of the stations early because we ran out of materials. The number of consumables needed for a station varies depending on the audience present and how long the station will be open. A way to save more consumables is to explicitly teach the volunteers how much to use during their training so that they do not use too much. Another strategy is to switch the activity partway through ChemFest or to have a backup demonstration in case supplies run out. Another key lesson is that the exhibits need to be varied. Because families tend to stay for the entirety of the event, it is important to have different demos at each of the stage shows. We have also needed to increase the scope and level of activities each year. We noticed that many of the same families return each year, so we needed to make certain that the demonstrations evolve along with the audience.

With respect to training SAACS and AXΣ members, our main focus has been to ensure that events are performed safely, but we have realized that more effort should be directed toward explaining the science behind the demos for all age groups. Finally, we ask for audience feedback every year after ChemFest. Useful suggestions have included name tags for participating college students, flyers to advertise JMU Chemistry, and a presentation to elementary school students to inform them that college is an exciting possibility for all of them.

Skeleton Festival

Skeleton Festival is a 2-hour event associated with Halloween and the Day of the Dead in Harrisonburg, VA (*6*). Skeleton Festival is a public event organized by the City of Harrisonburg. Skeleton Festival was originally called Halloween on the Square and has been held annually since 2004 (*6, 7*). The City of Harrisonburg changed the name to Skeleton Festival in 2016 to help it become more inclusive.

Because Skeleton Festival is a city event, many local companies and organizations are invited to set up booths and tables. SAACS was first invited to participate in 2011 and set up a table to teach children about chemistry using hands-on demonstrations. Originally, Skeleton Festival was held from 10:00 a.m. to 2:00 p.m., but SAACS had a table only during the activities portion of the event that ended at noon. In 2016, Skeleton Festival was changed to a 6-hour event and lasted from 3:00 p.m. to 9:00 p.m. However, in 2017 it was shortened back to a

2-hour activity period. SAACS is invited every year to hold a table at Skeleton Festival during the activity period.

Every year, the SAACS demonstration coordinator plans the demonstrations and gathers volunteers, which include AXΣ members. The number of volunteers varies as the number of hands-on activities change. Each hands-on exhibit needs two or three volunteers during the event with a few extra volunteers around to help where an extra hand is needed, such as handing out and collecting safety goggles or gathering more supplies. All of the demos at Skeleton Festival are hands-on and interactive for the general public. Typically, they are geared toward a younger audience, mainly elementary-school-aged children, but people of all ages find them interesting. For the past two years, some of the demonstrations have been repeated, like shaving cream art and silly putty (4). In 2016, we introduced a mushroom cloud demonstration and CO_2 bubbles; using dry ice for both (4). In 2017, there was water bending with static electricity on a balloon, balloon skewers, and edible gummy worms (4). All of the demonstrations are designed to be safe enough for the general public to conduct them.

We strive to improve our Skeleton Festival experience each year. A surprising lesson we have learned is that there can be too many volunteers. Even though the event is outside (so there is space around the tables), there have not been enough hands-on activities to keep every volunteer busy, and many volunteers are not involved in the demonstrations. However, it is difficult to turn them away, especially when Skeleton Festival may be their only chance to participate in a demonstration during the fall semester since it is on a weekend. (Students in SAACS and AXΣ are required to participate in one outreach event per semester.) Furthermore, we do not want to discourage people from volunteering for future events by turning them away from this one. Nevertheless, we may institute a cap to maximize each volunteer's engagement with the public.

Another area with which we are constantly concerned is safety. For example, we have enough goggles so that all children watching and all participating have a pair. We also try to use common household items for the hands-on activities, so that parents know the demonstration is fairly safe. The problem with household items is that hands-on demos require a lot of consumables, so planning has to be done carefully to make sure the supplies do not run out before the demonstration ends. Over time, we have become more methodical with our planning, and we attempt to rotate the activities over the period of the demonstration to try to minimize the consumables used and to make supplies last longer.

Smaller Demonstrations

SAACS and AXΣ also participate in many smaller demonstrations throughout the year. These can range from elementary school science, technology, engineering, and math (STEM) days to after-school program activities to freshmen chemistry orientation. In addition to engaging the public, these events are also used to keep SAACS and AXΣ members active in their respective organizations. SAACS and AXΣ work together to provide volunteers for the shows, and SAACS leads periodic training workshops to ensure all students are capable of running the currently popular demos.

Typically, SAACS is asked to organize smaller demos 1-2one to two months prior to the event. The first step is choosing what will be performed, which varies based on the event. If it is a STEM day, generally the event includes three to five hands-on activities that are continuously happening for two to four hours. For stage shows, the performance will include five to six demonstrations that take a total of 30 minutes to perform. The smaller events need five to eight volunteers to run smoothly and two or three people to set up. The consumables used during these experiments vary based on the activity, but it is easier to estimate what we need since we often know the number of students in the audience. Similar to ChemFest, it is a good idea to have a backup activity if materials are completely used before the end of the event.

Some schools have also asked us to make crossword puzzles or relate the demos to information that the students are learning. Crossword puzzles help familiarize the students with the science vocabulary that we work into our show. We can also pair activities with Virginia state standards of learning (4).

The smaller demonstrations may incorporate interactive chemistry, or they may be done as performances to educate and entertain an audience. Events held at elementary and middle school STEM days and James Madison University's Expanding Your Horizons (8) are interactive, smaller demonstrations. These have included static electricity rice krispies and balloons, preparation of root beer and liquid nitrogen ice cream floats, golden pennies, and shaving cream art (4).

During March 2018, SAACS was asked to create a chemistry workshop for middle school and high school girls who are interested in exploring or entering STEM fields. The theme of the workshop was "Chemistry in the Home", which was very relevant and interesting to the students because the demos in which they participated used supplies that are found in their home and that can be purchased from their local grocery store. These activities included acid candy, silver-gold pennies, fruit volcanoes, milk swirls, and root beer floats prepared with root beer and liquid nitrogen ice cream made on site.

JMU Chemistry Freshmen Majors' Orientation and Second Home Child Care Center demonstrations are examples of events that are performed as shows. These events include demos such as exploding gummy bears, liquid nitrogen clouds, elephant toothpaste, and glowing pickles (4).

Our biggest concern for the smaller demo shows has always been safety. As previously mentioned, we have purchased a large number of goggles of various sizes to hand out during these performances. (Very small goggles, Gumballs StarLite SM, can be purchased from Gateway Safety online (9).) Fisher Scientific has also donated safety goggles for use during events. We also have become better at communicating our expectations of appropriate, safe behavior to our audiences. For example, during show demonstrations, liquid nitrogen-based demos are very exciting, but students always want to play with or touch the nitrogen. To combat this desire, we illustrate the danger of liquid nitrogen by immersing fruit and flowers into the liquid and then shattering them; we have found this to be particularly effective at curtailing this unsafe behavior. We also task a volunteer or faculty advisor with ensuring that the students nearest the stage remain seated.

We have also learned the importance of having a variety of outcomes or purposes for the demos we use. Originally, we repeated the same activities for most of the smaller shows and by returning to the same schools later, the students were seeing the same demonstrations many times. Therefore, we are always searching for new ideas to keep these events engaging for the volunteers leading them and for the students in the audience. As always, it is also important to make sure that we have an abundance of supplies for the demos that consume materials so that the functions do not have to be ended prematurely.

Conclusions

Based on our experience at JMU, it is possible to have two active cooperative chemistry organizations. Both SAACS and AXΣ have a long history at JMU, and they have not always had the good relationship that they enjoy today. By setting out simple ground rules, formalized at JMU in the Miller Accords (3), we have fostered an environment of mutual respect. Some of the most consequential rules are those that divide fund-raising efforts, stipulate leadership roles in each club, and provide for an independent arbitrator for disputes. Equally important, however, is our use of demos as a way to strengthen the bond between the organizations.

Students join AXΣ and SAACS because they love chemistry, and they share that love with the community through demonstrations. Therefore, we have worked hard to make sure that these events are not seen as belonging to just one organization, regardless of who organizes the show. SAACS students often recruit AXΣ members to help with demos like ChemFest and Skeleton Festival. We have also implemented an independent department outreach coordinator. The DOC helps maintain the neutral status of demos, helps us keep our demos fresh and safe, assists in institutional memory, and helps integrate our efforts with those of AXΣ and the entire Chemistry Department. By working together to reach out to our community, AXΣ and SAACS have come to see each other as valuable partners.

Acknowledgments

We would like to thank Drs. Donna S. Amenta, Thomas C. Devore, and Brycelyn M. Boardman for their oral history of the department and Dr. Iona Black for suggesting we contribute a chapter to this volume.

References

1. JMU Student Affiliates of the American Chemical Society Home Page. http://jmuacs.wixsite.com/saacs (accessed March 2018).
2. Alpha Chi Sigma Home Page. https://www.alphachisigma.org/ (accessed April 2018).
3. Miller Accords, 2011. Student Affiliates of the American Chemical Society. http://docs.wixstatic.com/ugd/b6bb4c_c2a3657b7b484037ac159fc2f43139 45.docx?dn=Miller%20Accords%202.17.11.docx (accessed June 2018).
4. JMU Chemistry Demonstration Page. http://sites.jmu.edu/chemdemo/ category/demo-database/ (accessed March 2018).

5. Landis, A. H; Davies, M. I.; Landis, L. "Magic Eraser" Flame Tests. *J. Chem. Educ.* **2009**, *86*, 577–578.
6. Brown, T. Skeleton Festival To Replace Halloween on the Square. *Daily News-Record*, Harrisonburg, Virginia, Oct 6, 2016. http://www. dnronline.com/news/harrisonburg/skeleton-festival-to-replace-halloween-on-the-square/article_ec537f4a-8c36-11e6-bcbd-6733964fddb1.html (accessed March 2018).
7. Harrisonburg Downtown Renaissance Presents 12th Annual Halloween on the Square, 2015. Visit Harrisonburg Virginia. http://www. visitharrisonburgva.com/wp-content/uploads/2017/05/10-16-15-Halloween-on-the-Square.pdf (accessed March 2018).
8. Expanding Your Horizons. http://www.jmu.edu/mathstat/eyh/ (accessed April 2018).
9. Gumballs Safety Glasses. http://gatewaysafety.com/products/eye/starlite-gumballs/ (accessed April 2018).

Chapter 4

A Conversation with the Presidents: Nurturing a Successful ACS Student Chapter

Verna J. Curfman, Quincy E. Dougherty, and Irvin J. Levy*

Department of Chemistry, Gordon College, 255 Grapevine Road, Wenham, Massachusetts 01984, United States
*E-mail: irv.levy@gordon.edu

The Gordon College Student Chapter of the American Chemical Society (ACS) is a vibrant community of students, with active members from the chemistry major as well as a number of other majors who have made a significant impact on our campus and on the community, writ large, with whom they have worked during the past 25 years. Like most chapters, there have been years of great activity and years where the chapter has lain fallow. Nonetheless, the past seven years have been extraordinarily productive through the leadership of a number of chapter officers and the rekindling of excitement brought by a new adviser. The chapter has engaged in outreach activities on our campus as well as in Boston, New York City, Washington, DC, and at ACS national meetings. Here, we present a conversation with the two most recent leaders who share their experiences in hopes of providing useful insights for other ACS Student Chapters.

Introduction

The Gordon College Student Chapter of the American Chemical Society (hereafter chapter) is a group that, amidst a 25-year history, went through a revival in 2011 with the encouragement of a new advisor, Dr. Joel Boyd (*1*), who mentored the chapter for three years. From this mentoring the chapter has flourished becoming a group of highly motivated and dedicated student members who have become an active part of the Gordon College campus and in the greater regional community. During the past seven years, the chapter has received national recognition as an outstanding or commendable chapter each year. Green chemistry is an important element of the departmental ethos at Gordon College and the chapter has also received green chemistry chapter recognition annually in each of the past seven years.

The chapter has become a winsome face of the Department of Chemistry during these years, frequently acting as a host to visitors—from our large annual Distinguished Green Chemistry Lecturer event, to a much-appreciated Science Carnival during homecoming, to outreach activities locally and beyond, and to hosting prospective students visiting the college.

Outreach activities are numerous; for example, the chapter has recently participated in the following activities on our campus:

- Hosts for Distinguished Green Chemistry Lecturer;
- Departmental hosts during campus visitation days for admissions;
- Science Carnival, for families attending Homecoming;
- Science Café, providing food and fellowship once a semester for students between morning and afternoon lab sessions; and
- Interchapter projects with Northeastern University and Bridgewater State University.

Our chapter has also participated in the following activities beyond our home campus:

- National Chemistry Week and Chemists Celebrate Earth Day outreach events at the Boston Children's Museum;
- National Chemistry Week and Chemists Celebrate Earth Day outreach events at the Boston Museum of Science;
- Successful student chapter posters at ACS national meetings;
- Local elementary school, middle school, and high school science, technology, engineering, and mathematics nights;
- Outreach to preschool- and elementary-age children in New York City schools (*2*);
- Oral presentations at ACS national meetings (*3–5*);
- ChemDemo Exchange Workshops at ACS national meetings;
- Attended Esselen Award lectures (*6*) at Harvard University;
- Monthly science club for elementary school children at Oliver Partnership School in Lawrence, MA;
- Weekly tutoring for high school students in Lynn, MA;
- Attended the Yale/UMass Boston Green Chemistry Symposium; and
- Green Chemistry outreach at the USA Science & Engineering Festival (*7*), Washington, DC, along with the partner organization, Beyond Benign (*8*).

Gordon College is a relatively small liberal arts college (about 1700 undergraduate students with three full-time faculty in chemistry, graduating about six chemistry majors annually); yet, as indicated from the list of activities above, we have students who are remarkably ambitious and talented. As a faith-based institution, we also have students who care profoundly about making a difference in the world and our department is intentional in our desire to empower students to begin to do so immediately, not just after graduation or later in their careers. We want our students to practice leadership skills and learn the rewards (and costs) of their leadership while they are with us. Consequently, the role of the chapter adviser at Gordon College is truly advisory, not to provide leadership per se. If the chapter cannot provide their own leadership, then our philosophy is that we are doing a disservice to provide it ourselves. There should never be a sense that the chapter is a noncredit course run by the adviser that the students "attend." Rather, the adviser empowers the students, acting as an ombudsman, cutting through the ever-present red tape to allow the chapter to realize its goals.

Since the chapter is truly student-led, it seemed wrong for the chapter adviser to write about our efforts. Consequently, the remainder of the chapter is an edited transcript of a conversation with our two most recent chapter presidents—Immediate Past President Verna Curfman and President Quincy Dougherty. We gathered at one of our favorite restaurants on a December afternoon to talk about the chapter and spent several hours trying to understand why we have been successful and what useful message we might be able to share with others to help improve their own chapters. Thus, we give you, "a conversation with the presidents."

Background of the Officers

Irv Levy (IL): Since we'll be reading your comments throughout this chapter, why don't we get to know you each a little bit? Where are you from? What is your major?

Quincy Dougherty (QD): I'm from Palm Beach Gardens, Florida. I came to Gordon College as a biology major thinking I wanted to go into medicine; then at the end of my freshman year, I switched to chemistry. I also have a business administration double major.

IL: What prompted the switch for you?

QD: After taking general chemistry and the first year of biology, I realized I liked chemistry a lot more than biology and then after looking at the upper-level courses that I would have had to take for both majors, I determined that I would be more interested in the chemistry courses than those in biology.

IL: Verna, same questions to you.

Verna Curfman (VC): I grew up in Volant, Pennsylvania, which is north of Pittsburgh.

IL: We are going to get that in the book!

VC: Yes! Volant, Pennsylvania—it's a town of 300 people. I came in as a biology major, pre-med, and that is what I left as, with a chemistry minor, too. I am currently applying to medical school and teaching as a lab instructor at Gordon and as a tutor for science subjects in a local school district.

IL: Who was the president prior to you, Verna?

VC: The president prior to me was Daruenie Andujar, a psychology major, who was also pre-med. So, we had somewhat of an unusual succession.

IL: Prior to that, it had been chemistry majors who were the leaders. Would you say that Daruenie had any special influence on the chapter? Did she tend to bring it in a different direction than the focus a chemistry or a biology major would have?

VC: I know that the draw for her was the education portion and working a lot with children. And I think that she brought in a lot of outreach to the chapter, not that it wasn't something that we didn't do before, but I think for her that was a really large goal. And she also was good at seeing the needs of the students, in that she really worked to help look at REUs [Research Experience for Undergraduates] and how to apply for those. We also had a resume session when she was president.

Role of Adviser and Chapter Officers

IL: Let's begin with a full-on honesty question. I am a very hands-off adviser; I frequently get surprised by what I learn that you are doing. My philosophy has always been: this isn't a course, it's not a requirement, it's nothing you need to do; I want it to be totally owned by you folks. I don't want people to say, "I'm doing Professor Levy's club." I want it to be your chapter, and I've always seen my role as an adviser to knock down roadblocks. If somebody says, "Oh, you can't do that," then I say, "Yes, they can." If you've got a reasonable idea that you want to pursue, then I will support you at any level with the administration to say, "Let them do what they want to do," but I don't tell you what to do and I don't actually spend much time thinking about what it is you're going to do. So, I'm just wondering: there are pluses and minuses to that, correct? How do you see the role of what you need from an adviser to be successful?

VC: Personally, I love hands-off. The time I thought about it the most was as we were working with a chapter at another university. We had to work really hard to set up meetings through their adviser, and when we got to this meeting the adviser left and they're just sitting in silence, and it was awkward because it was a teleconference. Then, it was just us leading and drawing out every question, which wasn't working that well. I was able to meet up with those folks in San Francisco during the ACS national meeting. After meeting with their officers, we realized that their adviser said no to them all the time. Basically, everything that they did had to go through their adviser for approval. They were telling me that they had this great outreach idea but their adviser ended up saying no because the adviser didn't have enough time to do the organizing for it. They were flabbergasted that I didn't need to ask permission to do anything.

IL: So, did you steal their idea?

VC: [Laughs.] We should have; it was a great idea! But, it was just so obvious to me how much freedom we had, in the sense that with that freedom we had the opportunity to do so much more than what you, as our adviser, had the time to handle.

QD: I have definitely enjoyed it as well. I don't think that I would have liked to be micromanaged and told what to do. I liked having the freedom to be able to do what we wanted and take on certain projects—maybe not do others—and that we could make the chapter what we wanted it to be.

VC: I also think this was a really good way to develop leadership, because it wasn't that you weren't there to support us, but to be successful we had to rely on ourselves. We had to build people around us to get that support, or try to find people, instead of just going and running to "the parents" for help. It generally wasn't something that crossed my mind; only when there were tasks that we couldn't do administratively. Then, that was when you were helpful.

QD: I also think that it was a really great experience to plan all the events and activities ourselves. If we had an adviser who just planned everything and told us where to go and what to do, I would definitely not have grown as much as a leader.

IL: I already mentioned I'm a hands-off adviser, so let me explain a little bit of my history to understand why I am this way. I entered college in 1976. I got really involved in the chemistry department. I joined ACS as a student to be able to get discounted subscriptions to journals, and somehow or another found out that you could have a student chapter and our school didn't have one. It didn't exist, and I thought, "Oh, I should do this." I was always a person who liked to make things happen. I was that way in high school and again in college, and it never occurred to me that an adviser would be anything except a signature so that I had permission to do what I wanted to do for my chapter. When I became an adviser, it never occurred to me to think that my role should be anything but a signature and to be there when you had questions or ideas you wanted to run past me. Also, one of my favorite roles is to be the town crier of your successes. I am very proud of what our chapter does, and so I put it up on our Web site and I announce when you folks win an award and the entire faculty finds out about it. And it's also my job to be your signature on authorizations when you need money. I've never wanted it to be "Professor Levy's Club"; instead, I want it to be "our chapter," where each leadership team brings their own personal flair to it.

I do worry sometimes that I'm too uninvolved in your day-to-day work. That when you have challenges, you may almost have a feeling that you shouldn't bother me about it. But I *am* your adviser. We ran into a little of that this past fall, right? There were many times when you didn't have enough people to engage in the activities, so everything kept falling on the leaders.

QD: Yes, I don't like asking for help with anything really; it's hard for me to say, "Hey, I need help with this," so I'd just rather do it on my own and try to make it work. That's what I was trying to do this fall, so it wasn't that I wish you're more involved, or that you knew, or that you didn't, it was just more a case that this is my issue and I'm going to handle it.

IL: I know you were very proactive to understand what people wanted and to let them know that their opinions mattered and that did help. So, does that mean as an adviser to you that my role should be to check in with you somewhat more frequently to ask, "How is everything going? Do you need anything from me?" Or does that just become more of a burden?

QD: I think maybe that could be helpful.

Chapter Leadership Continuity

IL: One of the policies built into the bylaws of our chapter is that we have officer elections at mid-year and this is unusual. This is something that Dr. Boyd brought to us when he helped to revive the chapter seven years ago. Most other chapters have elections in the fall, and the officers serve for the academic year. I wonder how the mid-year election might help or hinder your transition and the transition to another president? Verna, you've had a transition and you then handed off to Quincy. How did that help being transitioned in mid-year, and how did it work being transitioned out mid-year? And why do we do that?

VC: Well, I thought it was really helpful. I also had an advantage in that I was an officer before I became president, so that was also a huge help because I knew how everything worked beforehand. But since we have mid-year elections, the past president didn't just evaporate off the face of the earth; if we had any questions about how to contact people, how to accomplish some task, then they were still there. And I hope that was also helpful to you, Quincy, in that I didn't evaporate. The only time it has ever been a problem is that we confuse our student government on campus a lot!

QD: Yes, your name is still on the board of club officers in the student center!

IL: So, everything on the campus is based on an academic year instead of a calendar year. Was it good or irrelevant that your last semester of your senior year you were no longer an officer?

VC: Oh, it was really good that I wasn't an officer. Just with the logistics of being ready to graduate, with all the tasks that I had to do, it was good. It was also good in that people weren't directly relying upon me before I left. So that responsibility began to transition.

IL: Responsibility for tasks like the annual report, for example, but you were still very involved?

VC: Right. I didn't just leave Quincy to fend for herself; I could send her documents and help her prepare for it, but in the end, she was the one who submitted our annual report.

IL: Quincy, did you use that support?

QD: Yes, definitely. I relied a lot on Verna's notes and what she wrote up on the activities from the year that she was president, and I think that it would have been a lot harder to start from scratch, so it was really helpful to have that support and only have to write up half of the annual report.

IL: And not only the annual report, but just think if Verna was the leader and led very well until you graduated and then it's, "Bye-bye, everybody." Then, as a new president, Quincy would have taken the leadership in the fall, and even with the best of intentions Verna might not have been available when you needed something. It was also convenient that she was on campus and involved in the activities, such as the New York City trip (2), that she had planned in the fall. Verna mentioned that she had been an officer before; is that usually the way it goes, the president has been in an officer position, or is that necessary?

QD: I was not an officer before I became president. I think it might have been helpful, but I don't think it is necessary.

IL: Agreed, because obviously you've been very successful!

VC: Also, it's very dependent on the cohort of officers because with the cohort that was with my presidency, there was no one who was eligible to run for president.

IL: Why was that?

VC: We had four seniors and one student who was doing a study abroad.

IL: Oh correct, so when you have a student who is a senior, they are not going to be elected.

VC: Right, because they can't fulfill the whole time.

IL: So, it had nothing to do with criminal records?

VC: [Laughs.] We didn't do background checks.

QD: I also think that it's good for freshmen who are starting to get involved and want to be active to be able to run for an officer role as early as January of their freshman year. Then, they can remain involved and maybe become president eventually.

IL: So, since you weren't an officer previously, how were you involved as a member prior to becoming president?

QD: I was involved the semester before but was not involved during my freshman year.

IL: What prevented you, because you're obviously so involved now, what was it that made it not a high priority for you in your first year?

QD: I think that it was because I was a biology major and I didn't know as many people involved so I never really came to the meetings. Other members invited me and I said, "Oh, sure, I'll try it," but I never actually did.

Importance of Interdisciplinary Involvement and How Green Chemistry Helps

IL: Our chapter has a history of having lots of student involvement, including officers, who aren't chemistry majors. Why is that? What on earth makes somebody who is not a chemistry major want to be an officer in the Student Chapter of the American Chemical Society?

QD: I think that other majors want to be involved because we do a lot that is not just chemistry-related. A lot of the activities we do are focused on service and education and that appeals to more than just chemistry majors. It's also helpful because chemistry is a small department, so we need the support of other majors.

VC: I think it is important that we are so focused on outreach and supporting science education and also when we do so much with green chemistry, so many of the greater ideas behind the reasons of green chemistry are larger than just chemistry. I think that draws others.

IL: Yes, green chemistry is very catalytic in our success, I believe. Gordon College has been very involved in green chemistry and green chemistry outreach for more than a decade at this point. Our institution is a charter member of the Green Chemistry Commitment (9). And our students rapidly accepted and rallied around the notion of wanting to study chemistry in ways that are intentionally safer for human health and the environment (10). They recognize themselves at the front edge of this transformative time in the discipline, and they are true change agents, even from their first semester in college.

Can you describe some of the activities that the chapter has done these past few years?

VC: We host the annual green chemistry lecture on campus, bringing speakers from around the nation to our community. Recent speakers have been from the University of Oregon and the U.S. EPA [Environmental Protection Agency]. With elementary school children up through high school children, we do an E-Factor activity that demonstrates one of the principles of green chemistry using candy (*11*). We also run an activity where students create an auto paint using the principles of green chemistry, and then they paint a Matchbox® car (*12*). [Figure 1, below, was a gift to us from preschool students who made the auto paint and painted a memento for us.] When talking about renewable resources, we often do an activity where we create biodiesel from cooking oil; a very visual way to talk about sustainability (*13*).

Student Commitment

IL: I have always noticed students at Gordon, and probably everywhere, are overcommitted to all sorts of different activities that they can do. The people that you would want most active in your group are the people who are probably committed to a lot of other groups and, not that you wouldn't want others, but people who have a lot of enthusiasm and so forth probably already do a lot of other activities. So, what is the draw to do the chapter? What makes a student want to do that?

QD: Again, I think all of the outreach and education activities that we do draw in a lot of people. One of our officers told me that in high school she had to do community service hours and so when she came to college, she was looking for ways to continue to do community service and just couldn't really find them. Then, she found the chapter and saw it as a great opportunity to continue to serve and use her talents.

IL: Do you think that is unique to Gordon College—that there are students who actively want to do community service—or is every school like that?

VC: I think in a practical way, for people who want to go on to professional school, applicants typically need to have that sort of involvement. That is part of the draw, especially because it is something related to the scientific field and doing outreach at the same time. You can put a lot of what we do on your resume.

Departmental and Institutional Support for the Chapter

IL: Obviously, to do all that you do, you need support. Can you talk about different ways that the department and the college support what you do and maybe ways that we could do better?

QD: Financial support is really important. We get a lot of our funding from the GCSA [Gordon College Student Association]. This year, we applied for a fairly large budget and received most of it, which has allowed us to do all of the outreach activities that we want to do. We have never been unable to do something for financial reasons. Also, the chemistry department has been helpful as well, especially when we did some projects that came up in the middle of the year, after the budget was set. For example, the New York City outreach last year, which wasn't in our budget, required substantial support from the department.

VC: There are a lot of people in the department who are very willing to bend over backwards to help us in situations. Obviously, the funds we need are important but also the practical assistance when we need to accomplish some task—there are many who jump in to support us. We never feel like we're a bother.

IL: Do you feel that other organizations on campus have that kind of support? Or not? Is that unique? Or is it just a nice thing that you have that?

VC: In some ways it's unique, I think. I watched other groups be advised and the adviser says, "Oh, that's a nice idea," but I don't think that they always experience the desire or ability to do more, which is something that we get.

IL: Do you think that the fact that this group receives recognition nationally has helped give you the support because of your visibility, and does that have any difference from being just a regular club on campus?

QD: I don't think most students on campus know about our awards and what we do, other than the chemistry majors.

VC: I remember going to a GCSA meeting and they needed an update from us, and so I stood up there for 10 minutes telling them what all we had done, and they literally said, "Oh! Why don't we know about this?!"

QD: The same thing happened again this year. We had to give our midyear report, so we made this beautiful PowerPoint presentation and had pictures from our different events, and the GCSA leaders were all shocked at what we had done!

IL: That's amazing to me. I would think that you would be so well-known and the support would then be easier for you, but it sounds like you start over sharing your narrative every year. I can't even tell you how many times I've had someone from College Communications say, "So, what is this student chapter?" And I don't know where the disconnect is there, but it occurs to me right now that maybe—instead of a development officer, which is one of the roles in our current bylaws—we need to add a communications officer whose role is to constantly tell everybody how great you folks are and let the student paper know, let the student government know…

QD: Yes, that's what we are planning to do. We rewrote the roles for each officer this year to have someone responsible for recruitment and also someone responsible for public relations and getting the word out.

Chapter Meeting Management

IL: Our chapter is fun. We do really enjoyable activities, both on and off campus, and we travel places together and we go out to eat together—it's a fun group! How do you balance the desire to make sure that the chapter is fun and still, that it is developing professionalism? It's not just playtime; it's also about professionalism. How do you balance that; how do you work that?

47

VC: I remember thinking about this when I was leading because we'd have 20 people in a room and they'd all be talking very loudly over one another, playfully but not at all on task. And I was thinking, "Ok, how do we get this under control?" They're all peers and it's a leadership position where your peers have given you this position that you technically have no real power in. You have a ton of responsibility, but you have no hierarchy over them. And so, I remember thinking that because, especially my year, there were so many students the same age as me, and I remember having to start creating a very strict pattern for what we did in the meetings. For instance, we would have games at the end.

IL: Once they'd started playing, you'd lost control?

VC: Once the games started, you're fine, you have fun, but the meeting is over.

IL: You didn't need that as an icebreaker at the beginning?

VC: No, absolutely not because I knew that I could say, "Hey, we're going to have this time at the end; save your crazy and just listen to me for 15 minutes and then we can move on." But, it is hard; you don't want it to be too serious because we are young, it's part of the energy factor of it, but at the same time we have tasks that have to be addressed if we're going to accomplish anything.

IL: Right! If you have a teacher in New York who is anticipating that we are going to show up and be there for a really good reason, then you have to have had these meetings and work though the content.

VC: Part of that, too, is the pattern of having an officers meeting and then a group meeting. I don't remember who started that, or if it's part of the bylaws, but that was always very helpful because we'd have that core meeting, we'd get through all the topics we needed to talk about and prepare for, and then let the group know at our next full meeting.

QD: It was beneficial for me to have Verna's meeting format. We start each meeting with a fun icebreaker question, like, "What is your favorite element and why?", then we go into our business and have a game at the end.

IL: What kind of game?

QD: They range from, for instance, science Pictionary, to getting-to-know-you games, and lately we've done more themed games for the holidays. We did a Name That Tune for Christmas; just fun things like that.

VC: That would be the worst part of my week of planning—coming up with the game!

IL: So, to be clear, you meet weekly?

VC: Yes, officers do, every week.

QD: And the chapter meets every other week. So, we have an officers meeting, then a chapter meeting, and so on.

IL: Is it difficult finding time that's available for everybody with everybody's schedules?

QD: It was very difficult this year because all of the officers were very involved in a lot of other activities, so our normal meeting night did not work for us. We ended up doing an 8:30 p.m. meeting on Tuesday night, which is later than we wanted but it was the only time that worked for all of us. I think it discouraged some members from coming because it's later at night and also on a lab night for the biology students, so we had a lot of people in that lab who just didn't come to our meetings because they were so tired after lab.

Benefits and Costs of Chapter Leadership

IL: From your personal point of view, what is the best thing about being the leader of an outstanding chapter?

VC: I think it was being able to encourage the other chapter members. That was a really great part of it because I was able to go back to them after a successful event and say, "You did this!" I just remember those being such good moments, and I also think that it was a great way to encourage other people around me to do good things in maybe ways that were helping their professional development or helping our community. I think that was the most encouraging part of being the leader.

QD: I think that my favorite part of it is knowing the impact that we've had and that all of the amazing activities that we do are noticed and recognized. I think that's not only good for me to see that all my hard work is paying off, but that also for the chapter, as Verna said, it's nice to be able to encourage them through it all.

IL: What's the worst part about being a leader of an outstanding student chapter? Because it's not all perfect, right? There are costs associated with this, too; what are the challenges?

VC: Time management is a big challenge because it's a huge time commitment for the president, especially. As president, you definitely bear the brunt of the organizing. Balancing that was super hard for me.

QD: Also, one of our challenges this year was finding enough people to accomplish everything that we wanted to because we had all of these events we wanted to do, and there are some that we've done in the past that we wanted to continue doing, but we didn't have as many active members who were committed to do them. It is difficult to want to do so much but not have enough volunteers. So, instead of not doing certain activities, it came down to the officers stepping in, which led to some of the officers getting burnt out.

IL: That's part of the leadership challenge. How do you disperse the workload but you're in a confined group of however many you have to disperse it through, and in a short time—

VC: —And ensure that it gets done.

QD: Yes, that was my issue. I would delegate to the other officers and then it just wouldn't get done and I just said, "Ok, I'll do it."

IL: That's a tough one. You know, ACS has a Leadership Institute that offers a number of workshops for volunteers, such as "Leading Without Authority" (*14*). That's exactly what you're talking about. No ACS chapter president has the authority to say, "You must..." I can say that to my students because I can say, "You must..." or it's going to bring your grade down. Even then, they can choose to bring their grade down so I can't force them to do it, but I can put a lot of importance on it by having it pinned to a grade. But, none of the tasks you folks want to do allows you to say to chapter members, "You have to do this." Figuring

49

out how you have to bring that out is so important. Have you had successes of that sort or failures? This is something that every chapter runs into.

VC: I didn't do it perfectly when I was leading but part of it was just that I acknowledged that as their leader, I didn't have that authority over them in the sense that I never was one to use that tone of, "You must do this," but I also would use any respect that the people around me might have for me to prompt them to do what I needed to be done. So, a lot of the group was either a year older than me or my year and I would say, "I know that you have this free time right now. I would really appreciate it and it would make my life a lot easier if you would do this." There was a lot of being honest and being very personal. And I would have seven or eight members' schedules memorized at all times so that I could say, "I know you don't have class right now, so can you please go to the Accepted Students Day event for me?"

IL: [Laughs.] Wow. You were a stalker! So, you had their personal lives all figured out and in a schedule? That's remarkable but, knowing you, not surprising.

VC: It was all really, really well organized. But, the whole idea of being personable was really important in getting people to do what really needed to be done. Yet, it didn't always work because people could still say no, which is the issue. I think especially in the cohort before my presidency and during my presidency there were a lot of "yes" people, which was very good because they would say yes to me, they would do so much, and they would make the sacrifices. It would have been impossible to do it all on my own.

IL: How many people would you say that you knew you could depend on?

VC: Probably between a group of five and seven members.

IL: And the same question would go for this year, Quincy.

QD: Not that many.

IL: That's difficult... This notion of history—this organization has won awards seven years running—how does this history of what you've done in the past help and hinder?

QD: It was helpful for me to know the history of different events that we've done so I didn't have to start from scratch. There are some activities that we do every year. I could decide to continue those activities but also look for new ones. Maybe some of the hindering is feeling that we had to live up to past years' expectations. For example, we'd done tutoring off campus in the past so to let that go this year was kind of hard and a little disappointing.

VC: Agreed. Pretty much the same; it's a really nice model in the sense that you know what you could be doing but at the same time, there is the expectation and it's also sometimes a little harder to change if you're thinking, "This is the way that we've always done it." You know, that's always an issue. When you say, "I don't want to run this event anymore," some people will always say, "What? We've always done it this way."

QD: And I also think it's hard to live up to past chapters because with Verna's group there were a lot of involved seniors and then they all graduated so we had fewer actively involved members, but we were still trying to do the same number of activities, which was a little hard for us.

50

Big Hairy Audacious Goals

IL: I think an important motivator for any successful group is to have what management consultant, Jim Collins, refers to as a big hairy audacious goal (*15*). What is your big hairy audacious goal? What would you like to be able to see when you come back at your fifth-year reunion and find out about the ACS chapter?

QD: What I've been trying to leave behind are tools for the next set of officers to use; an operations manual. I've gone through and written up very specific details about what each officer should be responsible for, different tasks, and what needs to be done to accomplish them. I think that background information was a little lost this year. For example, we run a scholarship fundraiser that requires a lot of milestones to be met and even though we had the documents from the past, it was hard to know what exactly was supposed to be done and when to do it because it was all in different places. I'm trying to put it in one document so somebody can open it up and know exactly what is required to accomplish each specific task.

VC: I know for me, thinking about what I was leaving behind when I graduated was just knowing that chapter members were involved in something that was larger than themselves. Also, that reality was an idea that was important to me because I think that so much of what chapters can be about is the growth of each individual chapter member. If I could come back in five years and see that there is still an impact where students care about other people. That is important to me.

Final Thoughts

IL: So, do you have any other thoughts about what it means to be a leader of a successful ACS Student Chapter?

VC: I was thinking that chapters may have different definitions of success. When I have talked to other chapter officers at ACS national meetings, they are very quick to say, "We had 75 people show up for this event" and that's what they count as a success, while I didn't really ever see it as our numbers necessarily. That might have been as a result of being from a small school. I thought it was more the impact and the conversations that we had. I had a great conversation once with a Hamilton [Massachusetts] high school student who was embarrassed to tell me that she wanted to be a midwife. I was able to share my experience with this idea, and that it was a great field to go into, and to be an encouraging force. I think that idea of being able to affect one person at a time is at least as important as being able to say that we had a hundred people show up.

IL: And that impact doesn't just go one way, does it? What kind of impact has encouraging others through the chapter had on you and on the chapter?

VC: The impact of working with people throughout Massachusetts and the rest of the country has been almost indescribable for me and the other chapter members. When we were working in New York City last spring with elementary school children, we met a young girl called Zoë who painted a paper plate [Figure 1] as a gift for us when we were leaving. When Zoë handed it to me, she asked if I would always remember her. The truth is that she will probably forget who we were 15 years from now, but I will never forget Zoë. I know that she and all of the other people we've worked with have reminded me how a small deed can grow to have a larger impact. Overall, meeting people like this little girl has made our chapter continue to work hard toward seemingly small goals and projects that have built over time to have a greater impact.

Figure 1. Memento of our visit to Pono School, New York City.

IL: Is there anything that you did to help prepare chapter members to become leaders?

VC: When I was being prepared, there was very much a period that I was targeted to first attend activities, and then to have small responsibilities in them, and then suddenly I was helping to run those same activities, and then I was in charge of them. And so, especially being an officer before being the president helped with that. But, even before then, being able to identify that a certain person can handle an amount of responsibility and then giving them that responsibility was intentional. When I was president, I remember identifying three people as very good potential leaders, and they weren't even part of the chapter yet, I just knew of those people, one of whom was Quincy. I think it's important to identify others in whom you see potential and then invest time in them. I think you can't get anything out of someone without investing a little bit into them. I also want to watch out for the quiet ones, as both Quincy and I tend to be. We don't have to be the loudest person in the room, but we can take on leadership roles.

IL: So, if you knew in advance the time that this took and then you look at the end—you're out of it, Verna, and you're sort of in the midst of it Quincy—would this be the path or the club that you would have joined, or would it have been, say, volleyball instead?

VC: Oh, I wouldn't change the experience for anything. As I was reflecting when I was applying for medical school, this was one of the most impactful experiences that I had in all of college. It taught me so much about who I am; it taught me so much about how to interact with other people; and it also taught me so much about being part of a larger organization and taking those values and applying it to something. In a professional sense, I learned and gained so much but also on a personal level, I did, too.

QD: Yes, I really love being a part of it and as I was thinking about whether to run for president again, knowing how much of a time commitment it is, I asked myself whether I really wanted to do it again. It basically came down to the fact that I love this so much and I would be sad not being as involved so I am definitely going to do it again, knowing what I do now.

IL: It bears noting that last month you were elected to a second term as our president. Only one other president has served two terms in our chapter.

QD: But, I mean, I did consider not running again. I did step back and evaluate that this was a lot of time; was it worth it? And it was a yes.

IL: And that, I believe, is the perfect end for our conversation!

Acknowledgments

Many thanks are due to those who have supported our chapter over the years. We would be remiss to not cite Drs. Ronald and Christine MacTaylor who were the initial leaders of the chapter, known then as the Gordon College Chapter of the Student Affiliates of the ACS nearly 30 years ago. Their work set a firm foundation that continues to stand today. For example, it was under their leadership and efforts that the Homecoming Science Carnival was begun. It has been very gratifying over the years to see them return during homecoming with their own children, who then participated in the science carnival and then, in one of life's wonderful circles, to have one of those children help to run the science carnival when she was a student at Gordon College.

Also, as mentioned earlier, Dr. Joel Boyd brought new vitality to our chapter as well as many activities and procedures that continue today.

Kathy Swierzewski is a faithful supporter of the chapter, sharing ideas, practical advice, and much wonderful food to support the officers and members of the chapter. Many thanks!

Since the reboot of the chapter, we have been led by the following presidents, each of whom have put many hours and their own unique contributions into the chapter: Andrew Worth, Ben Stewart (two terms), Justin Andrews, Brittany Marshall, Daruenie Andujar, and our co-authors, Verna Curfman and Quincy Dougherty.

Funds to support the many efforts of the chapter have come from the GCSA and donors to the Gordon College Department of Chemistry.

References

1. Erskine College. Chemistry Overview Page. http://www.erskine.edu/ academics/majors/chemistry/ (accessed July 13, 2018).
2. Pono School and Thurgood Marshall Academy Lower School. Welcome to Pono. http://pono.nyc/ourjourney/ (accessed July 13, 2018). New York City Department of Education. Thurgood Marshall Academy Lower School Page. http://www.tmals.org/ (accessed July 13, 2018).
3. Curfman, V.; Ngo, I. CHED 83, How Student Chapters Can Have Impact on Green Chemistry and Social Justice Education. In *Green Chemistry: Theory & Practice*, Proceedings of the 253rd National Meeting of the American Chemical Society, San Francisco, CA, April 2–6, 2017.
4. Curfman, V. CHED 117, Recipe for Success: How Outreach Has Impacted Gordon College's ACS Student Chapter. In *Celebrating Chemistry through Outreach*, Proceedings of the 253rd National Meeting of the American Chemical Society, San Francisco, CA, April 2–6, 2017.
5. Dougherty, Q. CHED 94 (gc)^2: Gordon College's Commitment to Green Chemistry. In *Green Chemistry Student Chapters: Stories of Success*, Proceedings of the, 255th National Meeting of the American Chemical Society, New Orleans, LA, March 18–22, 2018.
6. Northeastern Section of the American Chemical Society. Gustavus John Esselen Award for Chemistry in the Public Interest Recipients Page. http://www.nesacs.org/awards/esselen/awards_esselen_recipients.html (accessed July 13, 2018).
7. USA Science & Engineering Festival Home Page. https://usascience festival.org/ (accessed July 13, 2018).
8. Beyond Benign Home Page. https://www.beyondbenign.org/ (accessed July 13, 2018).
9. Beyond Benign. The Green Chemistry Commitment Page. https://www. beyondbenign.org/he-green-chemistry-commitment/ (accessed July 13, 2018).
10. Anastas, P. T.; Warner, J. C. *Green Chemistry: Theory and Practice*; Oxford University Press: Oxford, 1998.
11. Beyond Benign. E-Factor Curriculum Page. https://www.beyondbenign.org/ lessons/e-factor/ (accessed July 13, 2018).
12. Beyond Benign. Science as Invention: Making Paint, 2010. NANOPDF Web site. https://nanopdf.com/download/doc-5ae7cb0a75f36_pdf (accessed July 13, 2018).
13. Beyond Benign. Synthesis of Biodiesel Curriculum Page. https:// www.beyondbenign.org/lessons/synthesis-of-biodiesel/ (accessed July 13, 2018).
14. American Chemical Society. Leading Without Authority Workshop. https://www.acs.org/content/acs/en/careers/leadership/descriptions/leading-without-authority.html (accessed July 13, 2018).
15. Collins, J. BHAG – Big Hairy Audacious Goal. Jim Collins Web site. http://www.jimcollins.com/article_topics/articles/BHAG.html (accessed July 13, 2018).

Chapter 5

Evolution of an AwESOME Chapter

Emily Cooper, Sadiq Shaik, Danielle Bautista, Ayan Ahmed,
Christian Gaetano, Meg Hockman, Sarah Kochanek, Ben Jagger,
Sarah Richards, Colin Schmucker, Liz Roeske, Emilee Renk,
Kimberly Daley, Jared Romeo, Matt Wilding, Ashley Biernesser,
Sean Noonan, Lauren Matosziuk, Ralph A. Wheeler,[1]
Nithya Vaidyanathan, Paul Johnson, Ellen S. Gawalt,
and Jeffrey D. Evanseck[*]

Department of Chemistry and Biochemistry, Duquesne University,
600 Forbes Avenue, Pittsburgh, Pennsylvania 15282, United States
[*]E-mail: evanseck@duq.edu
[1]Current address: Department of Chemistry and Biochemistry, Northern
Illinois University, 1425 W. Lincoln Hwy., DeKalb, Illinois 60115,
United States

Our American Chemical Society (ACS) student member
Chapter has a 65-year-old, rich history with a great number
of measurable successes. Over the years, we have found that
a meaningful American ChEmical SOciety Student MEmber
(AwESOME) chapter provides essential undergraduate
experiences in leadership, teamwork, and communication
to promote professional development and complement the
academic experience. Established in 1942, our chapter's
evolution occurred in two significant steps, each involving
substantial changes in vision and allocation of resources, in
1986 and 2008. This article focuses on the changes made during
the past decade, when we invested departmental resources and
realigned expectations to make our AwESOME Chapter a key
part of the Chemistry and Biochemistry Department's learning
environment. During that time, our chapter matured to become
a true team committed to engaging in the life of the department,
the university, and the community. It has become a source of fun
for the students and pride for the entire department. Our efforts
are described in three parts. The first section briefly presents a

historical framework of the chapter's beginnings. The second section outlines our trials and tribulations in transforming our 2008 chapter into what it is today. Finally, we describe our current state of affairs and the activities that establish our AwESOME Chapter as a premiere student organization of over 100 members with more than 20 major events per year. It is our intent to describe our efforts so that others may benefit from our experiences and be inspired to provide their undergraduates modern and necessary skills to compete in today's workforce.

Duquesne University

It is first prudent to place our efforts, resources, and chapter size into context. In this manner, readers may understand our evolution and decide if any of our strategies or events may be implemented at their institutions.

Duquesne University was originally founded and named as the Pittsburgh Catholic College by the Order of the Holy Spirit (Spiritians) in 1878. The university currently has 5998 undergraduates and 3258 graduate students, yielding a 14:1 student-to-faculty ratio. The Chemistry and Biochemistry Department was established soon after the university was founded, and its Ph.D. program was created in 1954. The department currently has more than 100 undergraduate majors and 35 Ph.D. students. Since 2003, with 16 tenure-track faculty, our external research funding has averaged nearly $1 million per year. Over the same time frame, the department has successfully acquired eight National Science Foundation (NSF) Major Research Instrumentation (MRI) grants and has continually hosted an NSF Research Experiences for Undergraduate (REU) site. Duquesne is unique in that it houses world-class instrumentation, facilities, and research programs similar to larger research universities, yet it offers undergraduates direct access to those facilities and our faculty is comparable to that of smaller liberal arts colleges. Our undergraduate students become involved in research typically by their sophomore year, and we maintain a culture of student service and engagement, attributes consistent with the university's mission.

Chapter History

Our American Chemical Society (ACS) Chapter was established on February 26, 1942, by Professor Hugh Muldoon, Chair of Chemistry and Biochemistry. The chapter started with 21 student members. The activities and impact of the early development of the chapter were not well documented. However, it is clear that the department made a deliberate decision to enhance and grow the chapter in 1986, when the chair of the department, Professor Andrew Glaid, recruited Dr. Theodore Weismann. Dr. Weismann had extensive network connections in the Pittsburgh region and was a research leader of Gulf Incorporated's geochemical division. One specific responsibility of Dr. Weismann's new position was to become the faculty mentor of the Duquesne Student Affiliates Chapter. In their first effort, Dr. Weismann led the chapter to partner with the Pittsburgh

Local ACS Section to initiate a regional undergraduate symposium in 1986, a mini-symposium that had a keynote speaker and student poster session. The annual event attracted circa 50 students and faculty, and the tradition continued for 32 years. Over this time period, the chapter's membership stayed relatively constant, with five to eight student members per year. Their efforts were rewarded by winning the Outstanding Chapter Award 13 times over 20 years, starting in 1988. In 2010, Professor Ralph Wheeler was hired as the Chair of Chemistry and Biochemistry. He recognized the need for a strong ACS student chapter, and orchestrated the changes to make chapter development a high priority. Chapter growth and development have been fortified as part of our tradition of undergraduate excellence under the leadership of Professor Ellen Gawalt, the current chair of the Department of Chemistry and Biochemistry.

Trials and Tribulations

The Department of Chemistry and Biochemistry underwent significant growth starting in 2000, including the addition of six new faculty members with more than a quadrupling of the number of chemistry and biochemistry majors over a 6-year period. The department's expansion necessitated a critical review of student programs, including the ACS Student Affiliates, now known as the Student Member Chapter of the ACS. Upon review, we found a number of positive features of the chapter. However, we concluded that significant opportunities were being missed with a greatly expanded faculty and student body, and that a strong student chapter that attracted and motivated students was no longer a part of Duquesne's undergraduate tradition. Specifically, we found that student participation and interest had faded significantly, a faculty member was carrying the majority of the load, and the chapter had become complacent, reclusive, and dated. Our interpretation was that the lack of student challenge, ownership, and autonomy were key factors that needed to be addressed. We decided to rebuild the chapter to be better aligned with the department's mission and provide essential undergraduate experiences in leadership, teamwork, and communication to promote professional development and complement a rigorous academic experience. It was clear that our 2008 effort was similar to starting a new ACS chapter, which provided an opportunity to create something special. We decided to realign and create a more modern, meaningful, and autonomous student organization that provided undergraduates the nontechnical skills and experiences not necessarily found in a rigorous academic program. We realized that such a restructuring would take considerable effort, resources, and commitment. Importantly, we took advantage of resources that were present at that time and did not try to force-fit solutions. A take-home message from our experience is that every institution of higher education is different and even though common problems with ACS chapters may exist, each institution has a diverse set of opportunities and resources at its disposal. Creativity, ingenuity, and patience are essential factors in addressing the small and large issues facing an ACS chapter. Support of the chair and department is essential for a chance of success.

Faculty Mentors

Our first and possibly most important decision for rebuilding the chapter was to assign the mentorship to a tenured, research-active faculty member, along with a faculty member who had been a co-adviser since 1988. The two mentors were a natural fit to establish a new, relevant, and modern chapter framework without losing touch with the successes and connections previously made. Both mentors actively encouraged students to take a more proactive and independent approach in growing the student organization. The department chair continues to be a key mentor in the process, since the students are required to form a professional and administrative relationship with the chair. It was important for us to establish that the chapter was more than a club and that it provided meaningful student experiences for professional growth and contribution to the university and its mission. The objective of the mentors was to challenge the student leaders with new and exciting events, and to build new skill sets in leadership, teamwork, and entrepreneurship, without overwhelming or infringing upon their academic pursuits. Over the first three academic years (2008–2011), the faculty member pair had to lead by example, be heavily involved in most aspects of the chapter, and, in many instances, initiate, complete, and oversee events and tasks. The amount of time and effort invested by the mentors and the students was substantial. However, the mentors' efforts were always in the background or from the sidelines, with the students receiving full credit. The investment in time and effort was necessary to build an operational framework, establish the chapter as a reliable and functional organization, create a reputation along with tradition, and instill student confidence and pride. The students' responses were excellent in each of the years, and they became confident and independent leaders. Due to the chapter's growth, the department's laboratory manager also became a mentor to assist with demonstrations, events, and fund-raising. The main takeaway is that keeping the chapter as a departmental priority by selecting the right faculty mentors to take charge with a clear vision, along with departmental support and recognition, is crucial to the chapter's long-term success.

Student Leadership

An advantage of assigning a faculty adviser who taught both organic and physical chemistry to junior and sophomore majors is that dependable students with natural leadership abilities could be identified to lead and build the chapter. In the first year, with only eight members, student interest in the chapter was modest at best. Initially, the best students were suspicious and viewed the chapter as a social club that had no real input into the few events offered each academic year. It was clear that to attract the high quality students the chapter sought, it was imperative to provide meaningful challenges for the students and allow them to take more ownership of the organization. It was relatively easy to identify talented juniors and after revealing the department's plan for the chapter revitalization, three individuals accepted the invitation to run for the elected positions. The students had a history of working well together and were able to self-sort on the available positions of president, vice president, and secretary. Each student ran for two positions so that the limited membership had choices. The election was held at the end of the spring semester. As soon as the other students witnessed the success and recognition of the chapter after the first year, student attitudes and motivation to join the chapter increased dramatically. The ACS student chapter started with eight students in 2008 and roughly doubled every year until we reached circa 100 members, which is nearly 100% of the chemistry and biochemistry majors. It was surprising how quickly momentum could be established for a student organization. The chapter grew rapidly, which required modified bylaws to include more student leadership roles.

It is difficult to overcome bad leadership at any evolutionary point of an ACS chapter. With a small amount of effort, leadership issues can be avoided. First, it is important for faculty mentors to oversee the recruitment and election process. Specifically, two of our faculty mentors teach both freshman chemistry and organic chemistry for majors, which is the pool of students who take leadership positions on the executive board. As a consequence of the smaller classes and intense interactions, the faculty mentors have an excellent perspective on the leadership potential and growth of each student in the cohort. Next, and equally as important, the current senior executive board members are charged with identifying freshmen and sophomores who engage proactively, meaningfully, and responsibly in activities and events that demonstrate the potential for good leadership over the year. We operate by merging the two lists, as each helps in providing additional background insight. The two lists typically agree on identifying those not to invite as junior executive board members, and frequently identify candidates who would have fallen through the cracks. Each student has a unique set of qualities that qualify them as potential leaders of the executive board. However, the three minimum qualities involve students who are in good academic standing, mature students who have earned the respect of their cohort, and students who demonstrate good time-management skills. Before we enacted our policy of faculty- and student-led identification of prospects, we had a few students in leadership roles who were ineffective. To remedy the situation, we either turned to other members of the executive board or the faculty member had to step in to assist. Neither solution is ideal and we have not experienced this situation for over a decade. Over the years, we have found that an informed decision on identifying and encouraging possible candidates for the chapter's election to the executive board is a crucial factor in chapter growth each year as well as overall. The need for strong student leadership should not be underestimated.

Continuity of Leadership

Despite the chapter's success in its first three years, the lessons learned did not translate from one cohort to the next. As anticipated, the senior executive board members were consumed with their new chapter responsibilities and academic commitments, but surprisingly did not communicate with the elected students who followed them. In retrospect, the separation should not have puzzled the faculty mentors. However, the lack of communication made the faculty mentors feel like they were "reinventing the wheel" each year. It was clear that the cyclic process caused unnecessary duplication of efforts and limited the chapter's growth. An important role of the mentors in building the chapter from year to year is to guide each cohort past previous challenges or issues. In response, an effort was made to build a stronger conduit between consecutive chapter executive boards so that the same problems were not being identified and solved year after year. In the following three years (2011–2014), a group of about eight juniors was selected earlier than before, near the midpoint of the spring semester. The eight students were chosen so that any one of them would be suitable to hold any position on the executive board for the following year. Importantly, the students were invited to the last month of weekly executive board meetings overseen by the seniors and were asked to shadow the seniors on the mini-symposium that was the last important event of the academic year. Having juniors attend the meetings, mostly as observers, allowed for the transmission of some information from the executive boards. The juniors were evaluated as potential leaders, and this gave them time to see how they could contribute to the chapter and which position they would run for in the election.. This model was only one step in the right direction. The continuity and exchange of information and experiences between cohorts remained inadequate. Subsequent executive board members started to realize the need for training in their junior year and took significant steps to modify the chapter's procedures for the training and preparation of successive cohorts.

Layered Leadership

Our Executive Board model changed in 2015 toward the standard ACS model of *elect*, *current*, and *past* variations on each executive board position. The complete ACS procedure was not fully adopted but, in a significant step forward, circa eight sophomores at the end of the spring semester were identified and invited to attend the executive board meetings during their junior year to become fully integrated in ACS chapter leadership. The students were identified independently by faculty mentors and current senior members of the executive board, as previously discussed. The students—rising juniors—became part of the executive committee in a nonelected fashion and were referred to as "junior board members".

The last few chapter meetings of the year are significant, since the executive board-elect members lay out the schedule for summer meetings to construct a master plan for the following academic year. The master plan is a written document, complete with detailed and well-thought-out objectives and goals. In this manner, the juniors become part of the planning stage for the next academic year and see firsthand how the planning of events and activities takes place. Throughout the year, the juniors are treated as equals, except for voting, which has three significant advantages. First, the juniors receive a valuable year of experience, time to decide which position they would like to attain, and time to reflect on how to improve upon current efforts and events. Second, the burden of initiating, running, and overseeing more than twenty major events per academic year becomes more manageable with a doubling of the number of hands on deck. Finally, the elected senior executive board members gain valuable experience in the skill of delegating to the junior members so that they can focus more on vision and larger problems, while the junior members can share in the necessary experience and the details required for events. We have found that developing junior executive board members enhances the experiences of all board members, is necessary to avoid reinventing the wheel each year, and is essential in building the chapter's long-term success.

Regular, Frequent Meetings

Once the junior and senior executive board is elected and assembled, there are a few protocols that are followed in the operation of our chapter to enhance our chances of success greatly. First, the executive board meets for one hour on a weekly basis, at the same time and place, throughout the calendar year. Since the students are involved in summer research experiences, the chapter is able to meet during the summer months. It is important to keep the chapter high on the priority list for the faculty and student members; too much time between meetings allows for other activities to gain priority in a fast-moving and demanding academic environment.

The weekly meetings necessitate that progress be made and reported, which keeps the executive board engaged and productive. In addition, the executive board meets with the general assembly of members every other week. The meetings have a split format between value-added activities, such as workshops (e.g., resume writing or overview of research opportunities), and communication of plans and events. There are a great number of events throughout the year, and members are asked to contribute to at least three to remain in good standing. Meeting with the whole membership is essential to keep the group informed and engaged. Our practice of regular meetings ensures good communication and overall progress toward the chapter's stated objectives and goals. It also inspires freshmen and sophomores to become involved when opportunities arise.

The president leads the meetings, and the vice president formulates the agenda in consultation with the president. The agenda has evolved and is now partitioned into two sections of old and new business. The first section, old business, provides a format for wrapping up recently concluded events or activities but, most importantly, it also allows for a concise summary and open discussion on how to improve events for the following year. We have found this to be valuable in the transfer of knowledge from one cohort to the next, while simultaneously inspiring the junior members to improve upon and take ownership of events in the following year. The topics of new business are determined by the master plan (discussed below) and any other relevant issues that might arise. The president keeps everyone on track to discuss the agenda. This provides valuable experience in leading a professional meeting, where it is important to motivate and harness the efforts of talented participants yet keep the meeting on topic to achieve meaningful solutions.

Master Plan

The second protocol essential to our operations and the chapter's success is the creation of a written document stating our goals and objectives for the academic year, which the chapter refers to as the "master plan". The newly formed executive board meets at the end of the academic year to start the process of establishing the next master plan and continues to meet over the summer to finalize it. The first step in the development of the new master plan is for all executive board members to read the ACS chapter review from the previous year. The first summer meeting is devoted to discussing what the chapter achieved along with its successes and weaknesses. The executive board collectively decides upon how it can improve based on the review and experiences from the previous year. The second step in the development is to read previous year's master plan. The subsequent summer meetings focus on planning the events. Each event and the opportunity for a new event is vetted fully. We have found that the discussion continues to improve the chapter since events are either removed, modified, or added. Our current master plan is extensive, with several goals and supporting objectives in the five categories of service, professional development, chapter development, green chemistry, and budget, as defined by ACS. It has taken 10 years and a considerable amount of work to reach this point. When we started in 2008, we had only the mini-symposium as an important event that we could build upon. We decided early on not to overextend our efforts and to add only one or two events per year in a controlled fashion. This allowed the chapter to focus on the quality of event planning and execution. A deliberate effort was made to grow our events to populate the five different ACS categories uniformly. This allowed the chapter to grow in a balanced fashion. Not every event that we envisioned and implemented succeeded. The faculty mentors and executive board members need to be ready to learn from their mistakes and have the necessary flexibility to create and support meaningful activities that support the chapter's mission. We find that the master plan is an essential step for the students in defining a clear path forward to improve their chapter.

AwESOME Chapter

Our American ChEmical SOciety Student MEmber (AwESOME) Chapter took on a new vision and mission in 2008 to better serve its changing undergraduate population. Our chapter's new vision statement is "to foster the academic excellence and professional success of its members". The chapter's mission statement is "to create professional and networking opportunities that engage our undergraduate students, to serve as the centralized core of learning and activities in the Department of Chemistry and Biochemistry, and to make chemistry more accessible to students and the public".

Every year we define our chapter goal in our master plan, as discussed above. Our 2017–2018 Goals, in no particular order, are below. For the sake of brevity, only a few are expanded on in the following sections.

(1) Fund-raise an additional $3000 to build toward an endowed university account (Budget).

(2) Plan and implement the second "Speak Simply" event across the university for all students to refine speaking skills (Professional Development).

(3) Redesign the Mole Day celebration with engaging and creative competition activities (Chapter Development).

(4) Identify, invite, and host a fall departmental seminar speaker (Professional Development).

(5) Build relationships with other organizations for cosponsorship of events (Chapter Development).

(6) Continue established K–12 in-class workshops and research descriptions with the local Pittsburgh schools (Service).

(7) Build new National Chemistry Week activities in conjunction with the Pittsburgh Local Section of the ACS (Service).

(8) Design and implement events for members to develop professional skills for graduate school or industry (Professional Development).

(9) Continue to lead the planning, organization, and implementation of the Pennsylvania Junior Academy of Sciences (PJAS) (Service).

(10) Maintain at least three green chemistry activities, including a speaker and demonstrations (Green Chemistry).

(11) Prepare and instill leadership in members outside of the executive board (Chapter Development).

(12) Create a new field trip option (Chapter Development and Professional Development).

Service

Consistent with the National ACS Organization and Duquesne University's mission, our chapter places a strong emphasis on service and considers it one of the most important components of our chapter's identity. Key administrative support includes allowing students to use ACS service to count as formal university class credit for their service learning requirement. We strive to serve our classmates, professors, staff, and community through education and events. We participated in over 18 service events last year to educate the surrounding community of all ages about chemistry and its business applications and career opportunities. Our chapter's events are spread out throughout the academic year. Briefly, our principal events in service were held in conjunction with PJAS, The Allegheny County Sanitary Authority (ALCOSAN), The Ellis School, Oakland and Central Catholic high schools, The Carnegie Science Center, and Duquesne University to contribute to National Chemistry Week and Earth, Mole, and Pi Days. The majority of events have evolved into annual events, and each has garnered positive feedback from the constituents and partners served.

Pennsylvania Junior Academy of Science (PJAS)

PJAS is a statewide organization for middle and high school students (7–12 grade) designed to stimulate and promote interest in science, mathematics, and engineering among its members through active and engaging research projects and investigations. The Commonwealth of Pennsylvania is divided into 12 PJAS regions. Allegheny and Westmoreland Counties define Region 7, which serves more than 800 students and their families. Pittsburgh and Duquesne University are in Allegheny County. The PJAS competition, which occurs on the first Saturday of February, has been held every year since 1934 at local high schools. Each of the more than 800 students has 10 minutes to present their research orally in front of their peers and three judges. Each student has an official mentor, typically a teacher from their home institution. Students are evaluated and are eligible to receive a wide variety of awards. An extensive award ceremony takes place on the same day, and all the students and their extended families (more than 2000 people) attend to accept their awards from PJAS and surrounding industries. Students receiving a first-place award are invited to compete at the state level, at the Pennsylvania State University campus. Starting around 2005, first-place Region 7 award winners reported that the novelty of traveling and presenting at a large university for the state competition was stressful and negatively affected their scientific presentations. In addition, the PJAS Region 7 leadership observed that other regions hold their competitions at local universities to better prepare their students and reduce the "shock and awe" of a large university campus. In response, Duquesne's ACS Chapter, the Chemistry and Biochemistry Department, the Dean's Office, and PJAS Region 7 leadership started discussions in 2010 and formed a partnership in 2013. Our ACS Chapter had to invest considerable effort to prepare for the competition as a university event. The resources required were nontrivial as the event mandated a majority of the university's classrooms over a full day as well as the reservation of our largest ballroom to accommodate more than 3000 individuals coming to campus on a single day. The partnership was successful and since 2014 the competition has been held on Duquesne's campus to better prepare our regional students for state competition.

Since the PJAS Region 7 competition was established on campus, our ACS Chapter has maintained a leadership role in planning and executing the annual competition. First, the ACS Chapter and Chemistry and Biochemistry Department coordinates with the University Event Coordinator and the PJAS leadership to oversee, organize, plan, and implement the PJAS Region 7 event every year. The planning and meetings start at the beginning of the fall semester and initially required monthly meetings with representatives from all stakeholders. Second, the majority of our chapter's membership serve in one of many different volunteer capacities. The ACS general members volunteer as judges for the individual competitions in all fields of science, mathematics, and engineering. PJAS rules require 3 judges for each session of 10 students, which necessitates that there are $(800 \cdot 3)/10 = 240$ judges for 800 students. The all-day event is broken into morning and afternoon sessions, so circa 120 volunteer judges are needed if they serve for both sessions. The judges evaluate each student presentation based upon scientific thought, experimental methods, analytical approach, presentation, and judge's opinion. Our ACS general membership covers approximately a third of the judging needs. The other judges come from regional schools and industries. Third, the junior and senior executive board members work with the faculty mentors to make sure that all volunteers, including judges, pass Pennsylvania

Act 33 (Child Abuse) and Act 34 (Criminal History) clearances when working with children or volunteering with vulnerable populations. Considerable effort is expended in making sure that the more than 200 volunteers have the necessary training and credentials. Fourth, our chapter works closely with Duquesne and PJAS leadership to raise funds and coordinate the special awards session that gives over $5000 in cash awards to more than 50 student recipients. The most notable awards are the four full-tuition, four-year scholarships to Duquesne University. Prestigious and significant awards from Westinghouse, University of Pittsburgh, Carnegie Mellon University, and Duquesne's Provost and Dean are given as well. The executive board members assist in the presentations by announcing winners and handing out awards. Fifth, several of the chapter members serve as data entry and tabulation workers at the all-day event. The ability to transfer judges' scores to the awards session in less than an hour is critical to the success of the competition for both the morning and afternoon sessions. Six ACS members work under PJAS supervision to enter the data into their web-based software to determine the level of student awards. Finally, several of our members and faculty serve in a mentorship outreach program in which they mentor high school students to complete their PJAS research project. Our ACS members present possible research projects to interested high school students at Oakland Catholic, while other students meet with interested students at Central Catholic High School. From the outreach, we have successfully engaged two students who completed PJAS projects in Duquesne laboratories. Our efforts have been recognized by reversing the declining trend in student participation, which is now over 800 students, and by winning the Duquesne University Teamwork Staff award in 2016.

Impact

Since 2008, our AwESOME Chapter has evolved to become a premiere student organization at Duquesne University, where our service events are directly aligned with the university's mission. Service events provide the basis for our chapter's greatest accomplishments in two complementary fashions. First, our efforts and service events directly impact young students in the community. We have targeted students from kindergarten to college. Our efforts make science tangible to them, increasing their curiosity and interest in chemistry and science in general. Significantly, our personalized events reveal and underscore the real and attractive possibilities for a future career in chemistry or science. Second, our chapter activities provide our AwESOME Board members essential undergraduate experiences in leadership, teamwork, and communication to promote professional development to complement their rigorous academic experience.

Professional Development

The strength of our AwESOME Chapter lies in its membership. Therefore, the professional development of our members is of the utmost importance, as reflected in our vision. From freshmen to seniors, we assist our members in realizing the necessary skills for modern-day employment that will lead them to a successful scientific career. We offered over five professional development events last year to membership, to the university community, and to the Pittsburgh region. Briefly, our principal events in professional development include opportunities to (1) join, experience, and be elected to the executive board; (2) identify, invite, and host a departmental seminar speaker; (3) develop professional skills for graduate school or industry by participating in workshops given by academic and industrial speakers on interviewing techniques and how to apply for summer research internships; (4) travel, present scientific research, and network at the National ACS Spring meeting and our regional mini-symposium; and (5) plan, organize, carry out, and host Duquesne's "Speak Simply" event. Tangible departmental and college support includes rewarding members who fulfill minimum service obligations by funding their travel to present research results at the National ACS Spring meeting. Due to the successes and positive feedback, the activities listed above have evolved into annual events. The professional development for our chapter is extensive but in this report we elect to focus and provide further details on both our Speak Simply event and our regional mini-symposium, which have become two of our top professional development events to provide opportunities in leadership and responsibility through mentorship and event involvement.

Speak Simply

Many professionals, no matter their selected profession, get caught up in the nitty-gritty details and technical aspects of "what" they do at their jobs. Many do not take the time to reflect and create a simple and understandable framework on the "why" component of their efforts that makes sense to the general public. Having the ability to speak clearly and break down complicated concepts into simple statements is key to becoming a successful professional. No matter your chosen specialty, all can benefit from "speaking simply". In the spirit of preparing our students for a better and brighter future as they join the global workforce, our chapter implemented an event modeled after the National ACS competition of the same name that helps students develop effective and professional presentation skills by giving an understandable 2-minute oral presentation on their undergraduate research. By design, this contest helps students sharpen their communication skills and reminds them of the importance of speaking concisely and using language accessible to all, especially the nonexpert.

The Executive Board planned and carried out this event over the last 2 years, and our keynote speakers, Dr. Bill Carroll (2016) and Provost Bruce Bursten (2017), both past presidents of the National ACS organization, spoke at the university on the importance of speaking simply and effectively to the public. Undergraduates, graduate students, and faculty members across many disciplines on campus were present for both presentations. Due to time constraints of the 1-day competition in the first year, only 20 students were selected to present during the event from the submitted abstracts. Student presentations were limited to 2 minutes per student without the use of visual aids. Students presented to a panel of judges from a variety of disciplines, mostly from the liberal arts. Judges scored based on interest of the topic, ability to engage the audience, and making the topic of interest to the nonscience expert. Awards were given to the three participants with the highest score in the form of cash and a plaque with three additional honorable mentions. The executive board raised money from Duquesne's Honor College for the awards and recruited judges from the university's Honor Society, Phi Kappa Phi. A small reception was held after the event. Although it was the first year for the event, the feedback was overwhelmingly positive from everyone involved. Dr. Carroll was especially encouraging and provided essential feedback and direction to strengthen the event. The judges were impressed with the student speakers and faculty members in the audience stated that they intended to build the event into an extra credit opportunity for their classes next year. Dr. Carroll and the Pittsburgh Local ACS Section nominated the event for the ChemLuminary Awards for Outstanding Public Outreach Event Organized by a Student Group.

Based on the feedback from Dr. Carroll and others, the chapter held Speak Simply the following year, but with one major change: The event was split into two parts. The first, optional session gave students the opportunity to offer their 2-minute presentation and gain valuable feedback from a panel of judges before presenting at the official competition. In addition, the students who elected to be videotaped at the first, optional presentation became eligible for the Most Improved Award offered at the competition the following week. Undeniably, the feedback provided by the judges allowed students to improve their presentations greatly. The majority of the students capitalized upon the feedback as a learning experience and were able to present a sharp, clear, and understandable summary of their research for the nonexpert. The day of the competition started with Duquesne's Provost, Timothy Austin, whose linguistic expertise allowed him to speak on the importance of language in communication, followed by the keynote address from Provost Bruce Bursten of Worcester Polytechnic Institute. The university and the chapter were fortunate to have presenters who greatly engaged and inspired the students. Student feedback was outstanding. As with all symposia, it is essential to invite talented individuals who want to make a difference with the audience. The 2017 competition had 12 students, and the majority took advantage of garnering feedback from the first, optional phase of the competition. After presentations by Provosts Austin and Bursten, the competition started with the judges reviewing the videotaped presentation of the student followed by the "live" presentation for the competition. Dramatic improvement was observed in all cases. Overall, the Speak Simply competition received numerous accolades from everyone involved and has become one of the chapter's most visible and growing events that impacts the professional development of all students at Duquesne University.

Regional ACS Mini-Symposium

Our AwESOME Chapter has partnered with the Pittsburgh Local ACS Section since 1986 to organize a regional undergraduate symposium known as a "mini-symposium" for undergraduates in the Pittsburgh region. The symposium traditionally has a keynote speaker from Duquesne and a student poster session that attracts circa 50 undergraduates and faculty each year. The mini-symposium was the only event that we retained when reworking our chapter in 2008. Despite its success, an underlying issue was that only a single faculty member from Duquesne was invited to present their research at the annual meeting. It was our opinion that inviting more regional faculty to present would re-energize, diversify, and build a stronger undergraduate regional mini-symposium. In reworking the plans for the chapter in 2008, we decided to keep the original spirit of the mini-symposium. However, we were aware of several highly productive faculty from regional institutions of higher education that had established strong track records in undergraduate research and placement of their students in graduate programs. We changed the format of the mini-symposium to invite three external speakers to present their research for 30 minutes with time for questions. The executive board members took the lead in identifying potential speakers in consultation with the faculty mentors. Every attempt was made to identify potential speakers across a range of disciplines to increase the diversity and interest with the students. In general, two faculty members from academics and one scientist from industry have been invited each year. The spring semester has many competing events, such as Spring Break, Easter Break, the National ACS meeting, and final examinations. We have found that the second or third Saturday of April is usually the best fit for the region's schedule. The mini-symposium is an all-day event. Upon arrival, everyone checks in or registers and then is treated to a continental breakfast. Directly after the talks, all those who registered are able to eat a free lunch sponsored by the Pittsburgh Local Section. The student poster session follows lunch.

The executive board invites, arranges, and coordinates all aspects of the mini-symposium. After finishing the speaker list, the students advertise, collect, and publish student abstracts for the mini-symposium. The students had a website created to collect abstracts and have everyone register. The chapter has found that, after making the symposium more inclusive with three regional speakers, the number of student abstracts more than quadrupled. There are now more than 40 abstracts per year at the mini-symposium. The executive board arranges for faculty, post-doctoral, and graduate students to judge the posters. Standard criteria are given to each judge, with each poster judged by three different experts. During the judging time, the students are encouraged to network with different faculty and students. Cash prizes are awarded to the top five individuals who excel at presenting their research. Every effort is made to spread the awards out across the different institutions participating that day. To conclude the day, raffle prizes are given out as well as Visa cards for poster award recipients. The mini-symposium continues to be an important part of the education and research environment in the Pittsburgh region.

Impact

The benefit of full member immersion in our chapter has immensely helped our members grow professionally and continue their development as scientists and leaders. Our chapter presents students with opportunities to lead and participate to their fullest ability while providing opportunities for networking and personal growth. Our chapter is structured in a way that facilitates the growth of our members. The tradition of student-to-student communication and leadership helps foster the growth that will be undeniably valuable to future employers. In addition, our chapter found that attending scientific meetings was extremely effective for our members' professional development. During these meetings, members were able to gain valuable experience presenting their research and also networking with scientific professionals, especially during the ACS National Meeting. However, our chapter is most proud of our introduction of Speak Simply. This event is geared toward all undergraduates, university-wide, in developing effective public speaking skills when presenting their research. Students learn how to prepare carefully and think critically about their research. Their two-minute-or-less presentation is perfect for an elevator speech or to use in interviews for graduate programs or jobs in the chemical enterprise.

Conclusion

Our AwESOME Chapter has made a measurable difference with students, the university, and the Pittsburgh community. First, we have found that the structure of our chapter empowers students to become more aware, thoughtful in their decisions, and critical and independent in their thinking. Important to the chapter's success is that essential undergraduate experiences in leadership, teamwork, and communication have been ingrained in a pattern of behavior to promote professional development and complement the academic experience. Second, our chapter has evolved significantly by making wise decisions and benefiting from the generous investment of departmental and school resources. We find that any element of our chapter success is directly related to support from the Chair of Chemistry and Biochemistry, Dean of Natural and Environmental Sciences, and Provost of the University. Third, all stakeholders have been on the same page, whereby the chapter is not a club, but an untapped avenue for learning and growth typically not offered in a rigorous academic program. In other words, attempt to create an AwESOME chapter only if everyone takes the effort seriously. Fourth, patience is a virtue. It has taken us almost a decade to realign and create a more modern, meaningful, and autonomous student organization that provides undergraduates the skills and experiences for future success in graduate work and entry into the global workforce. Many initial ideas will fail and barriers to success will be discouraging. These are lessons to be learned and the experiences will give rise to greater opportunities. Fifth, we realize that one size does not fit all, and we have taken advantage of resources that were present at that time and did not try to force-fit solutions. Flexibility and ingenuity are key aspects in program development.

Most importantly, in creating an AwESOME chapter, we have uncovered a few essential tips that can translate to other ACS chapters.

(1) Make sure that your institution is committed to creating an AwESOME Chapter. The provost, dean, and chair need to embrace, contribute, and share the vision. Identify faculty mentors who can and want to make a difference with the students and organization, then support them in the

venture. This effort should substitute for another responsibility, not add to a growing and endless workflow.

(2) Have control in identifying the right students for leadership positions. The students need to be responsible and high academic achievers. The organization should not exist to replace academic productivity but add experiences to enhance a rigorous academic curriculum. Poor student leadership is extraordinarily difficult to overcome when maintaining a functioning and growing AwESOME chapter. The need for strong student leadership should not be underestimated.

(3) Devise a system that allows for the transmission of institutional memory from one cohort to the next. Our system of junior and senior executive board members has been found to be effective and eliminates the need to reinvent the wheel every year.

(4) The executive board should meet on a weekly basis. The meeting should be planned and have a detailed agenda that is discussed with resulting action items, assignments, and outcomes. The weekly meeting should not be time to gossip or eat pizza but to address the business of the chapter.

(5) Make sure the students articulate their aspirations and goals before the academic year begins. We accomplish this by creating a master plan each year. It is a written document that is referred to throughout the academic year and it guides our path when other distractions arise.

(6) Motivate your students with realistic challenges that can be incorporated into their academic schedules. Our policy was to start small and add one or two events per year, but always stress quality over quantity. Before you know it, you will have an AwESOME chapter!

We are happy to report that, even though change is inevitable, our AwESOME Chapter has evolved to become a key part of the Chemistry and Biochemistry Department's learning environment and a focal point for student engagement in the life of the department, the university, and the community. It has become a source of fun for our students and pride for the entire department. Thus, our students fulfill their core mission and make our AwESOME Chapter a distinctive feature of our departmental culture.

Acknowledgments

The chapter thanks the Chemistry and Biochemistry Department, the Bayer School of Natural and Environmental Sciences, and the Provost of Duquesne University for their generous and continued support. Many Duquesne students have contributed to the success of its AwESOME Chapter on the Bluff. Unfortunately, not all students can be listed, despite the significant contributions that they have made. However, the current and former AwESOME Chapter presidents and vice presidents since 2008 are recognized as authors on this manuscript. The student authors and faculty advisors are grateful to the National Science Foundation (NSF) and generous donors of Duquesne University. Emily Cooper, Meg Hockman, and Ben Jagger thank the NSF for the academic scholarships from the Scholarships for Science, Technology, Engineering, and Mathematics (NSF/S-STEM, grant DUE-1259941). Danielle Bautista, Sarah Richards, Liz Roeske, and Emilee Renk thank the Bayer Scholars Program and Duquesne University for the academic scholarships and research fellowships. Matt Wilding, Jared Romeo, Ayan Ahmed, Christian Gaetano, and Sarah Kochanek express their fullest appreciation to Patsy and Frank Deverse for their Scholarship and Fellowship Program. Finally, Christian Gaetano, Sarah Kochanek, Ben Jagger, Jared Romeo, Matt Wilding, Ashley Biernesser, and Lauren Matosziuk thank the John V. Crable family for their Scholarship and Fellowship Program.

Chapter 6

Developing a Science Demonstration Show Outreach Program

Elizabeth A. Raymond* and Steven R. Emory*

Department of Chemistry, Western Washington University, Bellingham, Washington 98225, United States
***E-mail: elizabeth.raymond@wwu.edu; steven.emory@wwu.edu**

In this chapter, we describe the ways in which Western Washington University's (WWU) Student Chapter (SC) of the American Chemical Society has developed an effective outreach program over the past decade using educational science demonstration shows. The SC annually organizes and presents numerous shows at schools and for other groups in the community. Using a process that emphasizes communication, safety, and assessment, the SC continues to improve its outreach program. These activities build a cohesive SC while addressing some of the educational needs in the local community. The SC's efforts also present science and chemistry in a positive perspective.

Introduction

Western Washington University (WWU) has established a successful and sustainable Student Chapter (SC) of the American Chemical Society (ACS). Western Washington University is a public regional university that has strong traditions of student involvement in service to the university, professional development for students, and outreach to the community. Western Washington University's SC offers experiences beyond traditional lectures, laboratory, and research that have a tremendous impact on students and the institution. These include, but are not limited to, increased student retention and further engagement of students in scholarly activity (*1, 2*).

At WWU, we have established a successful *and* sustainable chapter by creating a clear leadership structure, a strong and supported membership base, and a focused and needs-aligned chapter program (*3*). In this chapter, we focus on our educational science demonstration show program, which is an important component of the SC's overall program. Safety is of the utmost importance for all shows, and ACS provides numerous resources to help chapters plan and conduct safe demonstrations (*4–8*). In addition to serving the broader community, regular demonstration shows help bring the SC together for a common purpose. The shows serve as a member recruitment and retention tool as demonstrations are fun and engaging for volunteers. The SC typically presents 10–15 demonstration shows in an academic year, serving approximately 1,000 people. To make the SC's outreach efforts sustainable and successful, the SC has developed a system to organize, plan, and continually refine its demonstration show program (Figure 1) that is described in this chapter.

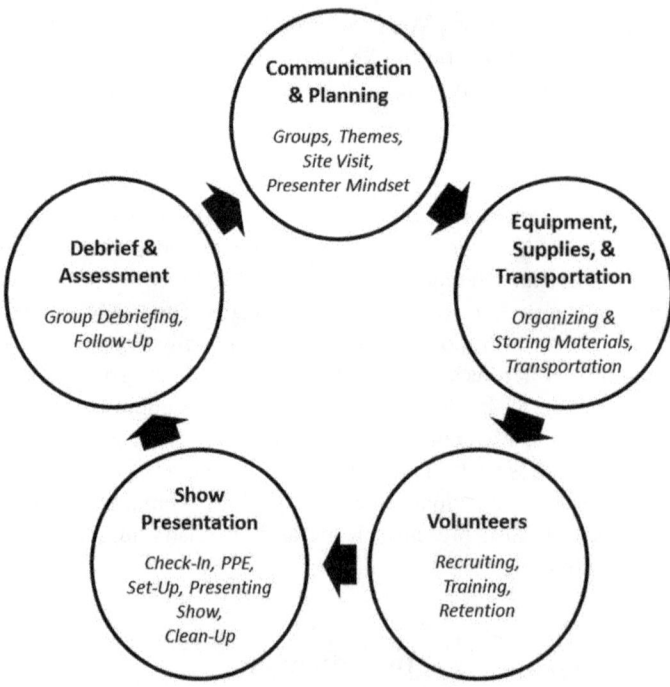

Figure 1. System to organize and refine demonstration show program

Communication and Planning with Schools and Groups

Groups

Due to the number of demonstration shows the SC performs (both on and off campus each year), the SC has become a popular option for school assemblies and community-based organization (e.g., Girl Scouts, Boy Scouts, and Girls Excelling in Math and Science) workshops. Several local elementary schools in the Bellingham Public School District regularly invite the SC to perform a demonstration show to kick off their science fair season. The SC also receives requests from parent–teacher organizations to do shows at schools not on the SC annual list. When the SC first started performing demonstration shows it contacted schools, particularly those elementary schools attended by children of WWU faculty. This made initial contact with the school much easier.

In Bellingham, schools typically schedule science fairs in the winter or early spring, making these times very busy for the SC. The SC finds it is easier to do one or two shows per week for a few months, rather than having shows more sparsely scheduled, for a couple of reasons. Shows are more successful if there are two faculty members and between three and seven student volunteers present. As both student and faculty schedules vary throughout the year, once several time periods are identified that work for the SC and local groups, the time and carpooling coordination becomes easier. Performing the same or similar shows at different venues in a short time span saves planning time. In addition, repeating shows in close temporal proximity provides excellent opportunities to evaluate and modify the show based on audience engagement and teacher feedback.

Demonstration and Theme Selection

Over many years, the SC has gradually acquired a large collection of materials and equipment for demonstrating different chemical and scientific concepts. Ideas for demonstrations have come from a variety of sources (9–12), and the SC is always looking for new ones to add to its repertoire. At least one chapter meeting each year is devoted to discussing interesting demonstrations SC members have seen in classes, on the Internet, or at a ChemDemo Exchange at an ACS National Meeting. From this assortment of new possibilities, the SC leadership decides which ones can be safely performed and discusses how they might fit into existing demonstration show themes.

Selecting a single scientific theme and a corresponding set of demonstrations to use at several shows allows the SC time to tailor each performance to its audience level. The SC typically performs at elementary schools, but while the demonstrations may be the same, how the SC discusses what students are seeing changes dramatically based on the level of the audience (e.g., a combined kindergarten through fifth grade audience versus a first grade only audience). Knowing and understanding the age range and current science topics of the audience is also important in selecting appropriate demonstrations. For example, if understanding the science underlying a demonstration requires knowledge of atoms and molecules, then it may not be as exciting (or even a significant learning experience) for younger students. The SC also considers whether demonstrations are going to be tempting for students to try at home. If they can be done safely (e.g., bubbles or cabbage juice indicator), then the SC provides audience members and teachers with instructions (both written and oral) on how to perform the experiment safely while also emphasizing the importance of adult supervision. If a demonstration poses an at-home risk and supplies are readily acquired (e.g., spraying alcohol salt solutions across open flames), then the SC will *not* perform this type of demonstration in a show. The safety issues and the potential for harm outside of the show must always be considered when selecting appropriate demonstrations.

Some of the SC's most successful shows have been developed after discussing science content in the different grade levels with teachers at local schools (*13*). Several years ago, the SC developed a "phases of matter" show to present to a set of first grade classes after they had finished their phases of matter learning unit. Knowing what the students had learned, the SC could comfortably discuss what the students were seeing using the vocabulary and concepts consistent with what they had learned in class. One piece of feedback from a teacher was that she really appreciated how much the SC did with gases, as they had been difficult for her to teach in a hands-on manner with her students. Knowing where students are in their science education helps make the show an informative and fun experience. Researching grade-level learning outcomes and science standards (e.g., Next Generation Science Standards) (*13*) allows the SC to meaningfully link demonstrations to what students are learning.

Preshow Site Visit

The SC finds a preshow site visit is essential before doing a performance at a venue for the first time. During this scouting trip, the SC has the opportunity to ask about the culture of assemblies at the school or group, see the facilities firsthand, and figure out the logistics of what can be done. For example, school assembly spaces often serve multiple purposes, such as gym, cafeteria, auditorium, or location for before- and after-school clubs. Because of this multipurpose nature, it is important to both see what the space looks like and find out when it will be available for setup and cleanup. For example, it is important to know if the cafeteria will be serving breakfast right up until the assembly starts, or if gym class starts immediately after the show ends. Figure 2 shows a typical checklist of factors to consider when looking at a presentation space. None of the factors listed prevent a show, but they certainly inform the selection of feasible demonstrations.

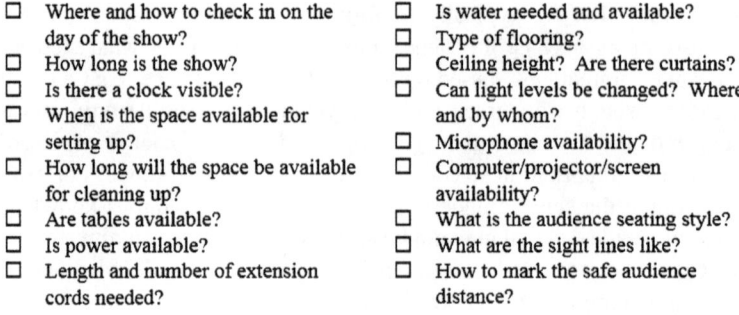

- ☐ Where and how to check in on the day of the show?
- ☐ How long is the show?
- ☐ Is there a clock visible?
- ☐ When is the space available for setting up?
- ☐ How long will the space be available for cleaning up?
- ☐ Are tables available?
- ☐ Is power available?
- ☐ Length and number of extension cords needed?

- ☐ Is water needed and available?
- ☐ Type of flooring?
- ☐ Ceiling height? Are there curtains?
- ☐ Can light levels be changed? Where, and by whom?
- ☐ Microphone availability?
- ☐ Computer/projector/screen availability?
- ☐ What is the audience seating style?
- ☐ What are the sight lines like?
- ☐ How to mark the safe audience distance?

Figure 2. Checklist of factors to consider at a presentation venue.

Meeting with someone from the school beforehand is also an excellent opportunity to ask about the assembly culture at the school. For example, the time allotted for a school assembly may be 30 min, but if that includes time for students to be seated and the principal to make a few announcements, then the SC may only have 20–25 min for the actual program. Demonstration shows are also very exciting for students, and they will often get quite loud and rather unruly. Asking if there is a school sign (e.g., raising a hand or a type of signal) that is used to bring everyone together to listen again is very important to know beforehand.

Appropriate Presenter Mindset

Faculty commonly think it is easy to speak in front of large audiences. While it may seem easy, the SC has found the need to very deliberately change the presentation style and content so that it is appropriate for the audience level. It is a common practice to ask college students to ask questions during lectures and presentations. Using this practice with an elementary school audience will guarantee the show does not make it through the first demonstration, because more than half the audience will ask *at least* one question. Asking open-response questions also sends a message that talking while a presenter is speaking is acceptable. The SC has found that asking very clear "raise your hand if..." style questions to be the most effective method for engaging the audience while maintaining decorum. Presenter directions such as "I want everyone to think quietly in their head about what would happen if..." rather than "What do you think will happen if...?" will get students in the audience to make a prediction without starting a room full of individual conversations. While it is important to learning that students articulate their ideas about science, a short demonstration show is not a very effective forum for this activity.

Equipment, Supplies, and Transportation for Demonstration Shows

Storing and Organizing Demonstration Materials

The SC is fortunate to have secure storage space for its extensive collection of demonstration materials in the chemistry building. This designated space allows the SC to store equipment and supplies organized by demonstration, rather than having to collect materials from different locations each time the SC prepares for a show. The SC stores *most* materials for each demonstration in transparent plastic storage bins with lids. The storage bin and laminated contents sheet for the "Elephant Toothpaste Demonstration" are shown in Figure 3. Having the materials organized in this manner makes preparing for a show very simple and quick. Once the SC decides on the demonstration "set list" for a show, volunteers just have to pull the correct bins from the shelves and load them into the transportation vehicle. A key to the success of this system is making sure that items are correctly put away and consumables are restocked after each use. This organization system is maintained by keeping laminated sheets of contents inside each bin. These sheets help both students and faculty easily and correctly put away supplies at the end of each show.

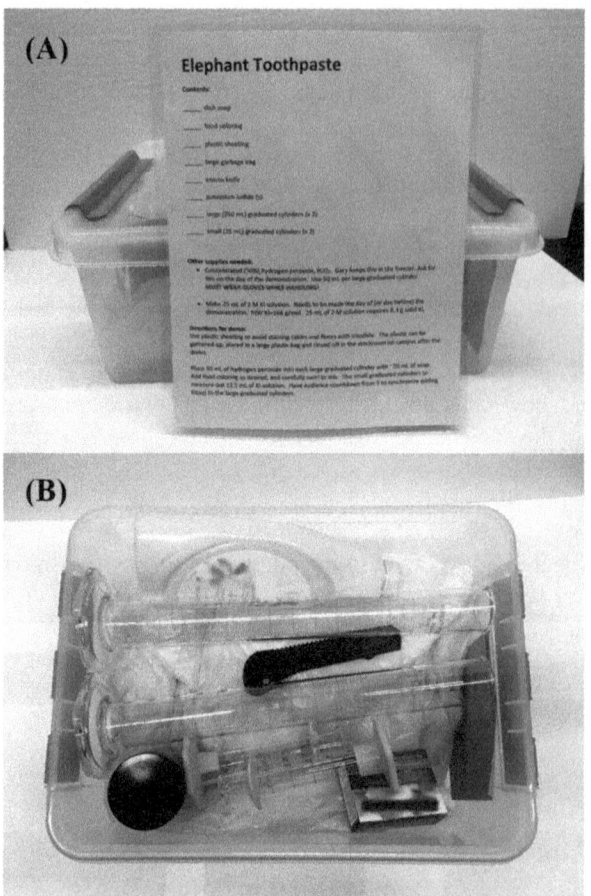

Figure 3. Storage bin system. (A) "Elephant Toothpaste" demonstration bin and laminated contents sheet. (B) Open "Elephant Toothpaste" demonstration bin. Note: concentrated hydrogen peroxide is not stored in the bin.

It is important to keep in mind when planning an off-campus demonstration show that the SC will not have the resources it is accustomed to on campus. This is obvious when thinking about the chemicals, supplies, and equipment needed for the actual demonstrations, but there are a number of contingency supplies that should be considered just as essential to have on hand. The list shown in Figure 4 is not exhaustive; the SC continually updates this supplies list.

Transportation to Venue

Transportation of equipment and supplies is another challenge when performing a show. Currently, a faculty advisor allows the SC to utilize their personal minivan to transport demonstration materials to and from a venue. The back seats of the minivan are removed and this provides adequate space for transporting demonstration equipment and supplies in a safe and secure manner.

☐ Fire extinguisher	☐ Container for transferring water
☐ Container for broken glass	☐ Old towel(s) for cleaning up water
☐ First aid kit	☐ Tape
☐ Acid/base spill kit	☐ Scissors
☐ Extra goggles and lab coats	☐ Extra lighter
☐ Gloves (bring a mix of sizes)	☐ Extra batteries
☐ Paper towels	☐ Tape/cones for marking seating areas
☐ Portable microphone	☐ Plastic garbage bag
☐ Extension cord(s)	☐ Plastic sheeting to protect floors
☐ Power strip(s)	

Figure 4. Checklist for contingency supplies for demonstration shows.

The bin organization system makes loading the vehicle simple. The demonstration materials needed for a 20- to 30-min show also easily fit in the trunk and backseat of two cars. Due to the relatively small size of the City of Bellingham (population ~80,000), student volunteers often carpool to shows. However, the time for transit and parking upon return to campus must be considered.

When planning shows, the SC carefully considers the safety of transporting chemicals (*6, 7*) and does not transport cylinders of flammable or compressed gases. The ACS provides safety resources that have proven valuable in helping to assess the safety and appropriateness of demonstrations (*4, 5*). The SC also observes all laws governing the safe transport of chemicals and consults the departmental safety officer for advice. When liquid nitrogen and/or dry ice are transported, these materials are always properly secured in the vehicle and vehicle windows are left open to avoid displacement of too much oxygen. Only the necessary amounts of chemicals needed to perform the demonstrations, not the entire stock, are transported (*7*). This helps to minimize cleanup in the case of spills or accidental container breakage. In all cases, chemicals require secondary containment, which in addition to being a safety requirement, makes packing the vehicle easier.

Volunteers for Shows

Recruiting Volunteers

Students have limited spare time, so they choose to spend their time on something they deem meaningful. Hence, a well-organized and effective science outreach program is an excellent recruiting tool for volunteers. In addition, openly advertising volunteer opportunities is important for creating an inclusive environment. Finally, personal invitations by SC leaders that encourage students to volunteer and participate in outreach activities have proven to be the most effective recruiting tool. Because volunteers are essential for a successful chapter, all SC leaders must play a role in recruiting and retaining volunteers. To help ease the workload, the SC appoints a volunteer coordinator whose responsibilities include e-mailing announcements about upcoming events and collecting a list of volunteers. The coordinator also sends out e-mail reminders that include the preshow meeting time and place and any special instructions.

Training Volunteers

Training volunteers to perform science demonstrations in a safe and effective manner is essential for a successful demonstration show program (5, 8). It is the official policy of WWU's SC that a *minimum* of one faculty advisor must be present at a show. This is done primarily as a safety measure, because a faculty member has the experience and authority to cancel a demonstration if conditions are not appropriate. A faculty member also has experience that can help handle an unforeseen situation or potential accident. The public speaking and classroom management skills a faculty advisor offers help maintain a high level of quality in the presentation.

The SC has attempted to schedule an annual demonstration-training day. However, this proved to be ineffective because not all volunteers could attend and there is often a significant delay between the training day and the shows. As a result, the SC conducts training for the demonstrations it will perform prior to every show. Training includes discussion of specific safety issues, procedures, and learning objectives. While this is not as efficient as a single training session, it is effective in that it covers the important material close to when the demonstration(s) will be conducted. Training focuses on the student volunteer(s) who will be conducting the demonstration, but other volunteers listen and can ask questions too. Only the demonstrations that will be performed (i.e., the "set list") at the specific show are covered. The SC maintains a list of what demonstrations student volunteers have been trained to perform. Safety information and procedures are again reviewed before the show commences at the venue.

Retaining Volunteers

To sustain an active demonstration show program, it is essential that volunteers be retained. Experienced volunteers are better able to perform the demonstrations in a way that engages the audience, and they become more cognizant of potential hazards. A fun and successful demonstration program helps retain volunteers, as they can sense the impact of their outreach efforts. In addition, SC leaders regularly organize postshow events (e.g., having ice cream) that express the SC's appreciation for its volunteers. These events also serve as social activities for the chapter and are used as debriefing sessions to review what occurred during the show.

Show Presentation

Prior to the day of the show, the SC prepares a demonstration "set list" (Figure 5) and the student volunteers review the safety protocols and procedures.

"States of Matter" Show Set List

Demo	Concept	Presenters	Time
1) CO_2 Bubbles • Bubble jar • Fish tank • Candles	$s \rightarrow g$ g g	Cassidy and Ash Stephanie and Doug Cassidy	8:30 – 8:37 am
2) Liquid Nitrogen (LN_2) Boiling	$l \rightarrow g$	Alena	8:37 – 8:39 am
3) Freezing & Thawing Balloons in LN_2	g	Carlos and Amanda	8:39 – 8:43 am
4) Bananas & Racquetballs in LN_2	s	Ash and Stephanie	8:43 – 8:47 am
5) Making Silly Putty	$l \rightarrow s$	Alena and Carlos	8:47 – 8:52 am
6) Whoosh Bottle	$l \rightarrow g$	Doug and Amanda	8:52 – 8:56 am
7) LN_2 Cloud	$l \rightarrow g$	Stephanie and Alena	8:56 – 9:00 am

Figure 5. Example of a demonstration show set list.

This preprepared set list is very important, as it helps to keep everyone on the same page and on task. It ensures that volunteers understand the expectations and are clear on how the show will proceed. In the past year, the SC has started saving electronic copies (e.g., photos or typed documents) of the set lists, which create a record of past shows. These records can serve as a starting point for planning future shows.

Check-in with School or Group

Once the SC presentation team arrives at the venue, SC leaders go to the main office and check in with the appropriate staff. It is important to do this early (if possible, 1–1.5 h before show time), as staff may need to unlock doors or show volunteers where tables and other equipment are located. This is also a good time to check the sound system and to review how to adjust lighting.

PPE Check for Volunteers

Before unloading equipment, volunteers are checked by a faculty advisor to ensure they have proper personal protective equipment (PPE) and clothing. Just in case, the chapter brings additional PPE (e.g., goggles and lab coats) that volunteers can use. If a student is still not properly attired, then he or she is not allowed to perform demonstrations that day. Instead, he or she may get to interact with the audience but will not handle demonstrations.

81

Demonstration Show Setup

Volunteers then unload the demonstration bins and bring them into the venue, and they help arrange tables and electrical cords. Electrical cords are minimized, taped to the floor, and left unplugged when not in use. Tripping is among the most common safety hazards during shows, and it can lead to potentially catastrophic accidents. To further ensure audience safety, a buffer zone is created using orange cones and tape placed on the floor. This ensures the audience remains a *minimum* of 10 ft away from the demonstration tables.

At this point, volunteers set up the demonstrations they were assigned to perform. Demonstration bins help improve the safety of this process, as they reduces foot traffic and the bins provide secondary containment for materials. Once all demonstrations are ready, a faculty advisor will walk through and discuss each demonstration as it occurs in the set list. The set list is printed and posted in several locations, so everyone knows what happens and when. Questions about the procedure, technique, and learning objectives are answered. The faculty advisor also reminds volunteers about specific safety procedures.

Demonstration Show

In addition to the volunteers performing the demonstrations, the SC assigns four individuals to special positions (Table 1) with important responsibilities at every show. It is the SC's experience that these positions are essential for ensuring a safe, fun, and effective demonstration show. During the show, one of the faculty advisors or an experienced student volunteer will serve as the master of ceremonies (MC). This person leads the show, and thus will use a microphone. Most schools have a sound system, but it is a good idea for a chapter to purchase a microphone and speaker of their own for backup. The MC first introduces themselves and the SC to the audience. The MC then typically asks presenters (e.g., student volunteers and additional faculty advisors) to introduce themselves. The MC presents each demonstration and leads the discussion of the science as the demonstration is performed. This allows the volunteers performing the demonstration to focus on performing it correctly. The MC asks the audience questions and gives the volunteers directions, depending upon the particular program. It is very important the MC know the "quiet" sign a group uses, as often members of the audience get very excited. Knowing how to use the quiet sign, and using it early, is important for setting a good tone for the show.

Ideally, the MC fosters interaction between the volunteers and the audience. The MC has preplanned questions for every demonstration and they ask additional questions throughout the show. Understanding the learning objectives is very important as it helps the MC guide the program. The MC also highlights safety precautions the volunteers take in performing the demonstrations. The PPE worn by all presenters is discussed, along with any demonstration specific safety measures (e.g., gloves, blast shield, or oven mitt). Integrating safety discussion throughout the show is important for establishing a safety culture for both the chapter and the general public (*8*).

Table 1. Assigned positions for demonstration shows

Position	Role	Responsibilities
MC	*Lead presenter and communicator for show*	• Introduce participants • Introduce demonstrations • Lead discussion • Adjust program timing as necessary • Wrap up discussion
Safety Monitor	*Watch for safety and demonstration technique issues*	• Bring attention to unsafe conditions • Help with technique problems
Timekeeper	*Ensure program timing*	• Watch clock and monitor progress of program • Alert volunteers for next demonstration • Communicate with MC about program timing
Demo Aide	*Help with unexpected issues and facilitate the show*	• Retrieve missing or additional items • Help with demonstrations as needed

In addition to guiding the audience through the science occurring in each demonstration, another important role of the MC is to help the audience envision themselves as future scientists. In all of the shows the SC performs, a clear emphasis is made that everything the audience sees is science, not magic! Every show is ended by asking the audience how many scientists they see in the room. After some audible counting, their responses typically correspond to the number of presenters. To this the MC responds that they have severely undercounted, pointing out that in fact the whole audience is comprised of scientists.

All participants in the demonstration program are responsible for safety. However, the SC has found that designating a safety monitor helps improve the shows. The safety monitor watches volunteers as they perform demonstrations and looks for safety and/or procedural problems. Most often this involves helping volunteers with non-safety issues such as technique. However, the safety monitor has prevented unnecessary spills and tripping hazards because they were focused on these issues. Typically a second faculty advisor or experienced student volunteer serves as safety monitor. The chapter has an excellent safety record, in part because of this important position.

The last two assigned positions are timekeeper and demo aide. The timekeeper is tasked with watching the clock and making sure the show is progressing as planned. If there is a time issue, they will inform the MC and the MC can adjust (i.e., shorten) the show to meet time constraints. The demo aide is designated to assist with any demonstration as needed. This may involve retrieving a necessary item, plugging in or unplugging an extension cord, or putting away a waste container. The demo aide is an important position that really helps the show run smoothly. It is also a good position for a new volunteer, as they can assist rather than lead a demonstration. The timekeeper and demo aide positions can be filled by a single person in cases of need.

Demonstration Show Cleanup

After the completion of a show, presenters need to be aware that some audience members may approach the demonstration area. It is important for presenters to watch out for possible unsafe situations. In general, it is best to wait until the entire audience has exited the venue before packing up materials. This ensures that presenters remain focused on audience safety. In cleaning up, tables and materials should be wiped down to remove any spilled chemicals from the surfaces. Equipment and materials should then be returned to the appropriate demonstration bin. Waste containers also need to collected, labeled, and properly sealed. Once this is complete, bins can be reloaded in the vehicle(s) for transport back to the university. Upon returning to campus, bins are unloaded, restocked, and stored in designated locations.

Show Debriefing and Assessment

After every activity the SC hosts, including demonstration shows, SC members and faculty advisors conduct a debriefing. It is through this debriefing process that the chapter has improved all of its programs over the years. In the case of shows, the participants first discuss what worked well. Did a particular demonstration engage the audience? What demonstration looked really "cool"? What demonstration was very enlightening? The group of presenters also discusses what did not work well. For example, were there any demonstration failures? If so, why did the demonstration fail? Can the demonstration be improved or fixed? Were there any demonstrations that were confusing for the audience? How can we improve the discussion? The group specifically discusses safety. Did anyone see any safety concerns or potential hazards that need to be addressed?

The faculty advisors also perform a follow-up with the leaders of the group (e.g., teachers, parent organizer, or principal) that coordinated with the SC. This is essential to make sure the show met the educational needs of the group. In particular, teachers can provide information about the classroom discussions that followed the program. This can be very informative about the effectiveness of demonstrations and the presentation. Any demonstration that was confusing can be brought to the SC's attention so that it can be addressed and improved for future shows.

We have found this process to be valuable, and the demonstration shows the SC performs now are different than those the SC performed 5 years ago. For example, the SC has removed some demonstrations that are too loud for indoor venues. We always use plastic tarps when using iodine to avoid staining floors. Demonstrations that produce odors are placed at the end of shows. When imploding metal cans, we always heat more than one can as a backup. Finally, we bring our own microphone and speaker, just in case.

Conclusion

Western Washington University's SC has continued to grow and evolve over the years. Fifteen years ago, the SC did not organize any demonstration shows. The outreach program started with a single elementary school assembly and now has grown into a vibrant, multishow program. It has taken time to develop the necessary systems that promote organization, self-assessment, and continual refinement. In the case of a demonstration show outreach program, organization is critical. Moreover, this allows the SC more time to tailor demonstration shows to better meet the educational needs of its audiences. Working together, WWU's SC has developed a successful and educational science demonstration show program that serves the community.

References

1. Graham, M. J.; Frederick, J.; Byars-Winston, A.; Hunter, A.-B.; Handelsman, J. Increasing Persistence of College Students in STEM. *Science* **2013**, *341*, 1455–1456.
2. Wyatt, L. G. Nontraditional Student Engagement: Increasing Adult Student Success and Retention. *J. Cont. Higher Educ.* **2011**, *59*, 10–20.
3. Emory, S. R.; Raymond, E. A. Key Components of a Successful and Sustainable Student Chapter at a Public-Regional University. *Building and Maintaining Award-Winning ACS Student Member Chapters Volume 1: Holistic Viewpoints*; Mio, M. J., Benvenuto, M. A., Eds.; ACS Symposium Series 1229; American Chemical Society: Washington, DC, 2016; pp 15–30.
4. ACS Chemical Safety Practices and Recommendations Homepage. American Chemical Society: Washington, DC, 2016. https://www.acs.org/content/acs/en/about/governance/committees/chemicalsafety/safety practices.html (accessed Jan. 17, 2018).
5. ACS Safety Guidelines for Chemical Demonstrations [Online]. Society Committee on Education, American Chemical Society: Washington, DC, 2018. https://www.acs.org/content/dam/acsorg/education/students/highschool/chemistryclubs/chemclub-demo-guidelines.pdf (accessed Jan. 17, 2018).
6. ACS Safe Transportation Recommendations for Chemicals Used in Demonstrations or Educational Activities [Online]. Society Committee on Chemical Safety, American Chemical Society: Washington, DC, 2014. https://www.acs.org/content/dam/acsorg/about/governance/committees/chemicalsafety/safetypractices/transporting-chemicals.pdf (accessed Jan. 17, 2018).
7. ACS Less is Better [Online]. Task Force on Laboratory Environment, Health, and Safety, American Chemical Society: Washington, DC, 2002.

https://www.acs.org/content/dam/acsorg/about/governance/committees/chemicalsafety/publications/less-is-better.pdf (accessed Jan. 17, 2018).

8. ACS Creating Safety Cultures in Academic Institutions. Safety Culture Task Force of the ACS Committee on Chemical Safety, American Chemical Society: Washington, DC, 2012. https://www.acs.org/content/dam/acsorg/about/governance/committees/chemicalsafety/academic-safety-culture-report.pdf (accessed Jan. 17, 2018).

9. Shakhashiri, B. K. *Chemical Demonstrations A Handbook for Teachers of Chemistry 1-5*; University of Wisconsin Press: Madison, 1983.

10. Flinn Scientific Chemical Demonstrations. https://www.flinnsci.com/resources/chemistry/chemical-demonstrations/ (accessed Jan. 16, 2018).

11. Steve Spangler Science Experiments. https://www.stevespanglerscience.com/lab/experiments/ (accessed Jan. 16, 2018).

12. Educational Innovations Lesson Ideas. https://www.teachersource.com/category/353 (accessed Jan. 16, 2018).

13. Next Generation Science Standards Homepage. https://www.nextgenscience.org/ (accessed Jan. 16, 2018).

Chapter 7

Location of an American Chemical Society Student Chapter: Balancing Service and Community Outreach between Urban and Suburban Settings

Theresa M. Dierker, Grace L. Nguyen, Justin A. Pothoof,
Danielle N. Maxwell, Cameron M. Johns, Kendra R. Evans, and
Matthew J. Mio*

Department of Chemistry & Biochemistry, University of Detroit Mercy,
4001 W. McNichols Road, Detroit, Michigan 48221-3038, United States
*E-mail: miomj@udmercy.edu

The location of the University of Detroit Mercy Chemistry Club allows for two major opportunities in conducting service and science outreach programs: following its mission in an urban setting and branching out into the suburbs. Club outreach events in these two environments exhibit specific differences and similarities. Herein, we describe multiple programs from both urban and suburban events with the aim of helping other American Chemical Society Student Chapters better define the influence of their location and how best to implement outreach events to harness the strength of their locale. We discuss details about events at urban and suburban sites, along with lessons learned relating to diverse populations of students, educators, and administrators.

Introduction

Goal 3 of the American Chemical Society's (ACS's) Strategic Plan for 2018 and Beyond is to support excellence in education (*1*). This goal is to be achieved by "fostering the development of innovative, relevant, and effective chemistry and chemistry-related education" (*1*). Key to the success of this goal is the component of coordinated, community-based, science outreach, whereby, through many different programs, including National Chemistry Week (2), Chemists Celebrate Earth Week (3), Kids & Chemistry (4), and Science Coaches (5), the ACS promotes chemistry in the public sphere.

At the University of Detroit Mercy, the Student Chapter of the ACS (Chemistry Club) has made excellent use of its location to reach out to both urban and suburban populations. The differences between these two populations in Southeastern Michigan are striking and have been this way for the past 60 years: Students who go to school in Metro Detroit urban areas are faced daily with crumbling facilities and infrastructure and dwindling resources per student (6, 7). In contrast, Metro Detroit suburban students, on average, have more modern facilities and plentiful resources per student. Interestingly, both populations suffer from the same budget-based problem of limited or decreasing extracurricular opportunities in science. The active members of Detroit Mercy's Chemistry Club have capitalized on their unique opportunity to serve both of these populations by partnering with multiple organizations and institutions, including the Detroit Local Section of the ACS. In the end, we have learned much about the particular needs of working with and connecting to diverse groups of students, educators, and administrators. In this book chapter, we detail our interactions with both urban and suburban groups so that other student chapters might glean information on how to best reach different audiences.

Urban Settings

The Detroit Mercy Chemistry Club is housed on the McNichols Campus in the University District of northwest Detroit. For more than 140 years, the university has pronounced its mission as "existing to provide excellent student-centered undergraduate and graduate education in an urban context" (8). A Catholic university in the Jesuit (Society of Jesus) and Mercy traditions, a Detroit Mercy education seeks to integrate the intellectual, spiritual, ethical, and social development of its students. The urban setting of the university is an integral part of every student's education at Detroit Mercy, and the work of the Chemistry Club is no exception. In fact, Detroit Mercy's core curriculum (9) specifically refers to coursework in the areas of cultural diversity, human difference, and social justice, where the urban backdrop of the university is emphasized as the perfect laboratory for exploring these integrating themes. Taken all together, these outcomes mesh well with the mission of the Chemistry Club: "As student members of the American Chemical Society, Detroit Mercy's Chemistry Club exists to promote chemical education through community service in an urban setting, provide preprofessional opportunities, and engage in fellowship through social activities."

When our Chemistry Club engages students, educators, and administrators in an urban setting, we have learned to anticipate three features for these interactions. First, all three of these groups are especially excited to host our demonstration shows or hands-on activities. As mentioned in the Introduction, the Detroit Public Schools Community District has been in a weakened state for decades. Many students, even when they enroll in science courses, have no chance to experience demonstrated chemical principles or to engage such principles in the first person. Second, a lack of on-site resources means that demonstrators must supply extension cords, tables, table covers, office equipment, and so on. The Detroit Mercy Chemistry Club has become accustomed to preparing for every eventuality in its visits, especially when the subject is chemical safety. For instance, many spaces large enough to contain a chemical demonstration show and audience may not be air-conditioned or even ventilated at all, necessitating the consideration of which demonstrations are appropriate for the setting and which ones are not. Finally, urban areas are more congested by automobiles, making the time required for travel, unloading of materials, and parking a more prominent concern. We have learned to budget for traffic, unpacking, and packing, making sure to allow for enough time for safe setup and rehearsal before a show. On the whole, our Chemistry Club has noticed that science outreach in urban settings differs from its suburban counterpart in important ways. What follows are descriptions of the more recent urban events in which we have taken part.

Girl and Boy Scout Merit Badge Clinics

The Detroit Mercy Chemistry Club joined efforts with the Detroit Local Section of the ACS to hold a large-scale event at which girls could earn the National Chemistry Week badge. During this event, sponsored by the Detroit Local Section and funded by a donation from General Motors, 200 Girl Scouts from around the Metro Detroit area gathered for a Saturday morning and afternoon on Belle Isle, a large community park in the Detroit River. The purpose of this annual event is to deliver the message that chemistry offers interesting and practical career options for women and minorities. The Girl Scouts rotated in three groups through several activities, including hands-on experiments, a presentation of women's careers in chemistry, a couple chemical demonstrations by our chapter members, and a canned food drive for a local food bank. Although the event was planned by the ACS Detroit Local Section, our student members were vital for staffing and organizing the activities, as well as for maintaining high enthusiasm among the attendees. The event resulted in a closer connection between the student members and the Local Section, and it served as an excellent opportunity to service the community through chemistry.

The Boy Scout merit badge clinic is an event held by the Chemistry Club every autumn, as a way for Boy Scouts to earn the Chemistry merit badge (*10*). The Chairman of the Chemistry Department at University of Detroit Mercy has been heavily involved in the Boy Scouts of America as a leader and helps connect the Chemistry Club with troops that would like to attend the event. Approximately 25 scouts between the ages of 10 and 15 years old attended our most recent merit badge clinic. To receive their Chemistry merit badge, the boys performed experiments such as chromatography with water-soluble markers and coffee filters and acid–base chemistry with borax and vinegar, and they designed Cartesian divers. The boys also used red cabbage, baking soda, and vinegar to experiment with "natural" indicator chemistry. The scouts enjoyed their time working in a college laboratory, and the volunteers had fun showing them the power of chemistry.

Technology Discovery Day

Every year, Detroit Mercy's College of Engineering and Science hosts a program for Metro Detroit high school students called "Technology Discovery Day." The purpose of the day is to expose students to technology and science in higher education. During the most recent Technology Discovery Day, our Chemistry Club members assisted with the setup of the gigantic event, volunteering to set up tables and chairs, check in students, decorate college buildings, and prepare for the hundreds of students who would attend. In addition, members performed several chemical demonstration shows. Many members of the Chemistry Club assisted with this well-attended event. For new club members, this event provided the first opportunity to participate in a chemical demonstration. We paired new members with veterans during the shows to run the demonstrations, which involved a burning dollar; liquid nitrogen; elephant toothpaste; and the crowd's favorite, hydrogen balloons.

Gesu Catholic Elementary School

As already explained in the Introduction, Detroit Mercy is situated in the University District neighborhood of Detroit, often referred to as "Northwest Detroit." Directly across the street from the university is Gesu Catholic Elementary School and Church (*11*). The parish was founded at the same time as the university by the Jesuit priests who made up the faculty of both. Currently, Gesu is a grade school enrolling kindergarten through 8th grade students. Every other year, the Detroit Mercy Chemistry Club visits Gesu School to perform chemical demonstrations in the school gymnasium. When these shows are performed, every member of the school community attends, including all students in grades K–8, staff, faculty, and administrators. Even as all who attend report back to us that they enjoy the demonstration show, the Detroit Mercy Chemistry Club also enjoys the event's proximity to campus and the chance to make connections through the shared Catholic mission of both of our learning institutions.

Detroit Area Pre-College Engineering Program

On five Saturday mornings during February and March, the Chemistry Club provides crucial teaching assistants for the Detroit Area Pre-College Engineering Program (*12*), which offers several different science and math courses to local middle school students. The program serves as an opportunity for young students in the Detroit area to experience laboratory work, which most of the participants are not able to practice at their current schools. During our most recent participation in this program, Club members assisted in teaching the course "The Chemical World" to middle school students. The lessons included chemical safety, electrochemistry, chromatography, acid-base chemistry, and heavy metal testing of water samples. Each Saturday program lasted for approximately 3 h. During that time, the students focused on one particular area of chemistry and diligently worked through the entire experimental process. A group of 20 students, paired with about five teaching assistants, created an excellent learning environment. The student members had a great time interacting with the young, enthusiastic chemists-in-training.

Harms Elementary After-School Program

In 2018, the Detroit Mercy Chemistry Club embarked on a new partnership with Harms Elementary School in the Mexicantown neighborhood of southwest Detroit (*13*). Members of the Club attended the after-school program and engaged approximately 30 elementary-aged students. First, our members helped the students with their reading skills by choosing story books to read aloud together. Next, our members served as teaching assistants for writing and math as the students completed their homework. At this point, the students were a little restless, so Chemistry Club members performed a short demonstration show to teach a few fundamental chemical principles. Finally, because it was a warm day, everyone present enjoyed a fresh batch of liquid nitrogen ice cream to cool off. Given that Harms Elementary School is more than 100 years old and has never seen any extensive renovation, the students at its after-school program were very appreciative of our visit, in terms of both the volunteering and chemical demonstration aspects.

Suburban Settings

More than five million people live in the Metro Detroit area, encompassing the populations of Wayne, Oakland, Macomb, and Washtenaw counties. More than four million of these people live in cities suburban to Detroit. Because the McNichols Campus headquarters of the Detroit Mercy Chemistry Club lies within the boundaries of the City of Detroit and because our mission speaks directly to our urban setting, outreach to the immediate area has always been our first priority. However, the Club is uniquely situated only 2 mi from the northern suburbs of Detroit, and campus sits right next to a major expressway connecting us to the western and southern suburbs as well. In the past 15 years, the Club has made a point of performing outreach in the suburban setting in addition to our main urban charge.

The Chemistry Club has found two key ways to branch out its science outreach efforts into the suburbs. The first is through the Detroit Local Section of the ACS. We have learned that many large-scale events (see the next section) need volunteers, which we can provide, in addition to planning our own demonstration shows to "piggyback" on these larger-audience events. Second, we have formed multiple partnerships with suburban Catholic school systems in need of science enrichment. Shrine Catholic Schools (*14*) and Our Lady of Sorrows Catholic School (*15*) both initially made contact with us, reasoning that our well-known demonstration shows could be made into "prizes" for their annual school auction fundraisers. Both schools listed our show as a lot for bidding, with the highest bid winning a private demonstration performance for their daughter's or son's homeroom class.

When our Chemistry Club engages students, educators, and administrators in a suburban setting, we have learned to anticipate two features for these interactions. Again, these groups are usually very excited to host our demonstration shows or hands-on activities. Students in even the more affluent suburban schools of Metro Detroit can have limited chances to experience demonstrated chemical principles or to engage such principles firsthand. In addition, we have found that suburban schools can be more concerned with events aligning with established curricular goals or specific learning outcomes. Knowing your audience is always essential, as there are certainly different kinds of chemical demonstrations for 8th graders versus kindergarteners! Taken on the whole, our Chemistry Club has noticed that science outreach in suburban settings differs from its urban counterpart in a few ways. What follows are descriptions of the more recent suburban events in which we have taken part.

Collaborations with the Detroit Local Section of the ACS: Cranbrook Institute of Science and the Detroit Zoo

Every year, the Detroit Local Section of the ACS holds a huge National Chemistry Week event at the Cranbrook Institute of Science, which is situated about 8 mi north of Detroit. At this event, Kids & Chemistry (4) volunteers provide hands-on demonstrations for both children and adults. The Detroit Mercy Chemistry Club supplies volunteers for the hands-on tables and also prepares multiple demonstration shows to be presented during the day. We have performed these duties with the members of the Chemistry Club from Grosse Pointe North (Grosse Pointe, MI) High School, who also prepare their own demonstrations. About 200–250 attendees total have been entertained by the Chemistry Club's demonstration shows at this event.

A little-known fact about the Detroit Zoo is that it is located in the suburbs of Metro Detroit, about 3 mi north of Detroit proper. The zoo is extremely family- and community-oriented. Compared to other city zoos, the Detroit Zoo is smaller, but this is a feature that gives the zoo its character. The size and location of the zoo, paired with its being open year-round, make it an excellent site for the Detroit Local Section of the ACS to hold its Chemists Celebrate Earth Week (3) activities. For more than 10 years, our Student Chapter has teamed up with the Detroit Local Section to participate in Kids & Chemistry (4) hands-on demonstrations at multiple zoo events. These meetings are a great way to get youngsters in the community involved in chemistry and science. Throughout the zoo, exhibits are presented both outdoors and indoors, mostly centered around ecological ideas such as green chemistry, water purification, and soil contamination.

Science, Technology, Engineering, and Math (STEM) Day Demonstrations at Shrine Catholic Grade School

In recent years, spurred by involved faculty, Shrine Catholic Schools has presented after-school activities for their students to gauge and inspire interest in certain topics regarding science. Thus, the Detroit Mercy Chemistry Club communicated our willingness to hold our own demonstration session in order to engage their interest in chemistry, specifically while talking to them about all the wonderful opportunities a degree path in science offers. This allowed us to inspire younger students and get them thinking about their future, and it also allowed our institution to do a little recruiting when we sent information home about what they did at school today! Events such as these touch base not only with the students we serve, but with their parents as well.

As a vehicle for our demonstration show, in 2014, 2016, and 2018, our Chemistry Club attended the Shrine Catholic Grade School STEM Days. Three chemical demonstration shows were performed for Shrine's preschool through 6th grade (more than 200 students). A Mission Microgrant from Detroit Mercy helped defray the cost of running the demonstrations, which can be significant, especially when the demonstrations are repeated multiple times during a single-day event (*16*, *17*). We performed the following demonstrations: "electric pickle" (in which an electric current is passed through an ordinary dill pickle until the sodium present in the brine begins to glow bright yellowish-orange); "giant's toothpaste" (in which the potassium iodide-catalyzed decomposition of hydrogen peroxide in liquid soap generates a very large volume of foam very quickly); "liquid nitrogen" (in which a number of household items are frozen in liquid nitrogen to show the difference between room temperature and extreme cold); and "helium and hydrogen balloons" (in which highly buoyant balloons separately filled with helium and hydrogen are ignited to show the difference between flammable and inert gases).

Our aim was to present amazing, thought-provoking, safe, and age-appropriate chemical demonstrations, and we achieved this goal as measured by a number of methods: (1) Anecdotally, the students, faculty, administrators, parent volunteers, and clergy present all enjoyed the show and left each presentation with a few tidbits of chemical theory. (2) All homeroom faculty (nine total) who brought their students to view the demonstrations were asked to complete a simple, four-question survey as an assessment. Teachers were asked to rate the safety, information level, usefulness, and professionalism of the viewed demonstrations. The results of these surveys were compiled and shared with the principal and the Chemistry Club members who volunteered to perform the demonstrations. The numerical results in each of these categories show that the educators polled rated us very highly in each of these categories. These assessment efforts are hoped to lead to better methods of demonstration performance and learning and teaching in the future.

At the outset, we had hoped that the connections forged through this event would serve to bring an awareness of Detroit Mercy to Shrine students, faculty, and parents. Also, during the events, we emphasized that Detroit Mercy is a great institution for studying STEM fields. We have confidence that the shared Catholic missions of Shrine and Detroit Mercy will provide an opening to future pedagogical interactions, including curriculum development consultation and Shrine STEM Club field trips to Detroit Mercy's College of Engineering and Science.

Conclusion

This chapter represents a synthesis of more than a decade of community outreach and service by the University of Detroit Mercy Chemistry Club. In our myriad interactions with entities both urban and suburban, we have noticed that the need for scientific or chemical education outreach has not decreased. The truth is quite the opposite. We have found that hands-on activities and demonstration performances, no matter what the setting, are sought-after supplements to slashed budgets and tight times in modern schooling. Through these interactions, we have learned a great deal about relating to diverse groups of students, educators, and administrators. We hope that the future of "Informal Science Education" (*18*) helps bring balance to the needs of urban and suburban populations and truly improves people's lives through the transformative power of chemistry.

References

1. Strategic Plan - American Chemical Society. acs.org/content/acs/en/about/strategicplan.html (accessed July 1, 2018).
2. National Chemistry Week (NCW) - American Chemical Society. acs.org/content/acs/en/education/outreach/ncw.html (accessed July 1, 2018).
3. Chemists Celebrate Earth Week - American Chemical Society. acs.org/content/acs/en/education/outreach/ccew.html (accessed July 1, 2018).
4. Volunteer with Kids & Chemistry - American Chemical Society. acs.org/content/acs/en/education/outreach/kidschemistry.html (accessed July 1, 2018).
5. ACS Science Coaches - American Chemical Society. acs.org/content/acs/en/education/outreach/science-coaches.html (accessed July 1, 2018).
6. Higgins, L. Detroit's Schools Score Worst in the Nation Again, but Vitti Vows That Will Change. *Detroit Free Press*, April 10, 2018. freep.com/story/news/education/2018/04/10/detroit-schools-again-worst-nation-rigorous-national-exam-while-michigan-overall-sees-no-significant/493893002 (accessed July 1, 2018).
7. Bosman, J. Crumbling, Destitute Schools Threaten Detroit's Recovery. *New York Times*, Jan 20, 2016. nytimes.com/2016/01/21/us/crumbling-destitute-schools-threaten-detroits-recovery.html (accessed July 1, 2018).
8. Mission & Vision | University of Detroit Mercy. udmercy.edu/about/mission-vision (accessed July 1, 2018).
9. Official List of Approved Core Curriculum Courses | University of Detroit Mercy. https://udmercy.libguides.com/shared_governance/approved_core_courses (accessed July 1, 2018).
10. Benvenuto, M. A.; Mio, M. J. Connections Between Service Learning, Public Outreach, Environmental Awareness and the Boy Scout Chemistry Merit Badge. In *Service Learning and Environmental Chemistry: Relevant Connections*; Roberts-Kirchhoff, E. S., Mio, M. J., Benvenuto, M. A., Eds.; ACS Symposium Series 1177; American Chemical Society: Washington, DC, 2015; Chapter 4, pp 67–72.
11. Gesu Catholic Church & School. gesudetroit.com (accessed July 1, 2018).

12. Detroit-Area Pre-College Engineering Program (DAPCEP). dapcep.org (accessed July 1, 2018).
13. Harms Elementary School. harms.schools.detroitk12.org (accessed July 1, 2018).
14. Shrine Catholic Schools in Royal Oaks, Michigan. shrineschools.com (accessed July 1, 2018).
15. Our Lady of Sorrows Catholic School. school.olsorrows.com (accessed July 1, 2018).
16. Mission Oriented. *Spiritus Magazine*, Spring 2016. udmercy.edu/faculty-staff/marcom/spiritus-archive/2016-04.php#book/9 (accessed July 1, 2018).
17. Royal Oak Shrine Hosts STEM Day Events. *Daily Tribune*, Feb 8, 2016. dailytribune.com/events/20160208/royal-oak-shrine-hosts-stem-day-events (accessed July 1, 2018).
18. Pratt, J. M.; Yezierski, E. J. Characterizing the Landscape: Collegiate Organizations' Chemistry Outreach Practices. *J. Chem. Educ.* **2018**, *95*, 7–16.

Chapter 8

Fund-Raising To Keep the SFU Chemistry Club Active and Engaged

Rose A. Clark, Hannah C. Schorr, and Edward P. Zovinka*

Department of Chemistry, Saint Francis University, Loretto, Pennsylvania
15940, United States
*E-mail: ezovinka@francis.edu

Chemistry clubs are only as good as the activites in which they become involved. To keep a club active and engaged requires not only student effort but also financial support. At Saint Francis University, our active program requires fund-raising to be a priority. We have used fund-raising as a fun way to challenge our students and get them involved in the lifelong learning endeavors of problem-solving, writing, and presenting their ideas. Students and faculty advisers have raised money by not only selling items, but also by writing proposals to the university, to American Chemical Society (ACS), and to foundations, as well as soliciting local companies. The challenge of fund-raising has driven our chemistry club to an ACS "Outstanding" and "Commendable" award-winning level.

Introduction

Saint Francis University (SFU), founded in 1847, is the oldest Franciscan institution of higher education in the United States. It is located in Loretto, PA, on a 600-acre southcentral Pennsylvania campus in the Appalachian Mountains. Currently, the university has a total enrollment of ~2700 students, with ~1750 undergraduates.

The Saint Francis University chemistry department offers a program approved by the American Chemical Society (ACS) that is staffed by eight PhD faculty with two masters level staff and services ~50 chemistry majors. As previously detailed, the SFU Chemistry Club engages in many service activities through the Rural Outreach Chemistry for Kids (ROCK) Program and other campus activities

(*1–3*). The ACS chemistry affiliates and the Gamma Sigma Epsilon (chemistry honor society) student leaders are excited to get involved, which leads to a need for funding. To keep our outreach activity level strong, efforts are made to write grants, work with the university advancement office to cultivate relationships with local companies, and take advantage of government funding opportunities.

Why Fund-Raise?

An important question the adviser needs to consider with the chemistry club officers and members is: what is the point of fund-raising? Your club may actually be required to hold a fund-raiser. For example, SFU requires all recognized campus organizations to fund-raise at least one time per year to receive money from the university. Even so, what will you do with the money?

Learning to fund-raise is a problem-solving endeavor, which is critical to teach our students. The club chooses an activity/outreach project it wants to accomplish. How can they make it happen? The group has to come up with the idea, justify the idea, write a proposal, and learn how to present their idea to others to convince them to fund the project. To be successful, chemistry club students must be engaged by faculty and guided through the grant-writing/fund-raising process.

Engagement Strategy

Engaging students in outreach activities is an important goal of any chemistry club. Once students have expressed an interest in chemistry, actively engaging them in a useful project (educational, fun) is a useful tool to retain them as scientists. A university chemistry club provides the vehicle to connect the students to chemistry outside of a strictly academic (classroom/research) role.

As the membership and leadership in the chemistry club change each year, student leaders' interests, priorities, and goals change the club's focus. As a faculty adviser, one has to adapt and roll with the new efforts. The adviser needs to provide guidance, suggestions about interesting options, and explain why the options are important, but ultimately, if the students do not want to do an activity, they will not participate and you are going to do it alone—not much fun and not the point! The students have to drive the club and take ownership.

The SFU Chemistry Club elects new officers each year as required by the university. It is important for the faculty adviser to meet with the new executive team after election to mentor them and to discuss expectations, including the possibility of writing proposals to fund activities. In January of each year, after the election of the new officers, the advisers and executive board meet to talk about past and future activities. It is important to cover what activities the officers want to continue, what activities they do *not* want to continue, and what activities they would like to add.

After deciding on an activity plan, fund-raising is discussed. Since the new officers, as undergraduates, do not have much (if any) experience grant writing or fund-raising, it is often up to the advisers to recommend funding sources. The advisers motivate the students by discussing their future career aspirations and how the grant writing process adds value to their resume (whether or not the proposal is successful). Electronic copies of previous SFU Chemistry Club proposals have been maintained as examples to ease student concerns about time requirements and to serve as examples of successful proposals. The advisers stress that each grant agency has its own guidelines in its Request For Proposal (RFP) that must be carefully followed. After the students write a draft, the advisers meet with the officers to go over the draft, with the RFP, to ensure the officers are following the guidelines.

At SFU, the students engage in a number of community service, recruiting, and research activities through the chemistry club. By requiring the students to write proposal drafts, edit proposal drafts, work with the community stakeholders, and evaluate proposals to fund their activities, the students gain valuable experience. Serving in a professional role as grant writer, the students sharpen their academic and vocational skills, often without thinking about the "lifelong learning" skills gained. However, the advisers need to guide the students to funding programs, motivate the students, and review their proposals prior to submission.

Chapter Activity Due to Fund-Raising

Most people think of fund-raising as a challenge—and it is, for many reasons. When starting out, pick the easiest ones first, raising your chapter's visibility locally. There are many other possibilities!

Community Building/Campus Visibility/Social Aspects

Often, when starting a new chemistry club or rebooting one, the focus is on building community, but fund-raising can be considered a side benefit. Selling merchandise is a traditional method used to raise club funds or outfit a team, and is attempted by many organizations such as athletic teams, church groups, and campus club organizations. T-shirt sales have been, and continue to be, a decent fund-raiser for the SFU Chemistry Club. Students wearing chemistry club T-shirts raise the visibility of the club on campus and the shirts can be used as a uniform when students lead off-campus or outreach events. Other items the SFU Chemistry Club has sold include chemistry-themed lanyards and pint glasses. Students like the lanyards and they often attach their keys and identification cards to them. Meanwhile, the pint glasses offer a respite from selling T-shirts semester after semester. However, given the small size of SFU, the chemistry sales market is small, limiting the profits to approximately $50–100 per sales drive.

A newer SFU Chemistry Club fund-raising activity is selling Krispy Kreme donuts to raise money for funding science programming in local schools. The program has raised our campus and community visibility, providing good public relations. The 2016 Chemistry Club officers came to us in the summer and complained that fund-raising efforts were too internally focused instead of on *serving* others. This was an exciting moment for the faculty advisers since

"Service to the Poor and the Needy" is a University goal. Being a Franciscan University, where "we engage in active service to the poor and to those with special needs", the **Don**ating to **U**nderfunded **T**eachers of **S**cience (DONUTS) grant helps us live our mission. The eight Goals of Franciscan Higher Education embody our educational efforts at Saint Francis University (*4*). The students have embraced the importance of serving others and, during the summer, the students created DONUTS with students selling Krispy Kreme donuts to fund grants to the Altoona–Johnstown Diocesan K–8 schools. The DONUTS fund-raising event has been held each of the last two fall semesters (Fall 2016 and Fall 2017). The club sold around 205 dozen donuts each year, raising over $700 each time. The money funds three $250 grants (with extra support from other sources). During the first year, the SFU Chemistry Club was also able to fund activities such as solving a mystery using chemistry techniques including chromatography, and learning about collecting and analyzing data through crystal growing.

Community Outreach

An important aspect of any chemistry club is reaching out to the community to demonstrate the fun and usefulness of chemistry (*3*). The SFU Chemistry Club actively leads over 200 K–12 outreach events each year and fund-raising is a necessary and continuing effort to make all of these events possible.

In 2003, the SFU Chemistry Club used an ACS Innovative Activities Grant to host a Boy Scouts of America Chemistry Merit Badge program. The Boy Scouts spent two different Saturday mornings earning their Chemistry Badges with funds used to pay for chemical supplies and lunches. After this initial event, the Chemistry Club worked with the SFU Alpha Phi Omega (APhiO) Service Fraternity to develop the SFU STEM University, an all-day event for the Boy Scouts. At this event the scouts select one merit badge to earn, spending a Saturday (8:30 a.m.–2:30 p.m.) completing the requirements. Scouts can choose merit badges such as Astronomy, Robotics, Nuclear Science, Animal Science, as well as Chemistry! By alternating hosting responsibilities with APhiO, the SFU Chemistry Club leveraged the initial 2003 event reaching 10 Boy Scouts to hosting 80 scouts for the 2017 STEM Merit Badge event. Additionally, because boys aged 12–17 are of prime interest to universities, we have found that the SFU Admissions and President's Office have been eager to help fund the program.

Since community service is an important component of the SFU mission, an ACS Community Interaction Grant intended to "improve the science learning experience of minority children through community interactions and projects" fits the SFU Chemistry Club perfectly (*5*). For example, in 2014 the club undertook a project in collaboration with the Gloria Gates Memorial Foundation (GGMF). The GGMF is a "non-profit organization that reaches children from low-income families. The goal of the GGMF is to encourage school performance and to build self-esteem and moral character so that youth will grow up to be contributing members of society by conducting an after-school program for children living in three of the low-income housing developments in Altoona, Pennsylvania" (*6*). The SFU Chemistry Club helped the GGMF move toward their goals. Using a Community Interaction Grant, the SFU Chemistry Club built a relationship with the GGMF and led hands-on science activities at all three low-income housing developments for over two years. The club visited a housing development once per month for the entire academic year—leading different science activities each time.

Recruiting Efforts

Departmental recruiting efforts provide an opportunity for the chemistry club to highlight the department to prospective students as well as interact with speakers and the departmental faculty. For the inaugural SFU Women in Chemistry Day in the fall of 2017, the SFU Chemistry Club used an ACS New Activities Grant to host two speakers and provide a unique learning opportunity for local high school chemistry students. In addition to touring the campus, high school students also toured the chemistry department, met with SFU students who discussed their undergraduate research efforts, listened to mentoring presentations by invited speakers, and participated in small group discussion sessions with the speakers and successful SFU chemistry alumni.

A variation on the Women in Chemistry Day was "A Day in the Life of a Chemist" undertaken in 1999 using an ACS Innovative Activities Grant. During the event, local high school students were invited to spend a day shadowing an SFU chemistry major, starting off at 8:00 a.m. and going until 3:30 p.m. (shortened for the high school students' schedules!). The matched pairs of students went to classes together, met with faculty, attended a first-year laboratory session, and participated in undergraduate research activities.

Both of these events have the added benefit of cultivating relationships with prospective students as well as with high school chemistry teachers. As high school chemistry teachers are an important recruiting connection, continuing efforts to cultivate relationships with the teachers are required as turnover through retirement, relocation, or job changes occurs frequently.

Research Funding

The chemistry club can also fund-raise for community-based research. Community-based club research, under the guidance of either the chemistry club adviser or some other faculty, provides an opportunity to teach research methods outside of the normal or expected arena. The efforts can lead to community engagement—both in the town and with the campus community.

In 2000, the chemistry club officers wrote a New Activities Grant proposal to ACS to study radon levels on the SFU Campus. The research was an eye-opening experience for the faculty adviser as well as the students as we discovered that not everyone is interested in data (either in collecting it or hearing about it) (7). The chemistry club members learned about the importance of Institutional Review Boards (IRB) by having to apply for IRB approval and also about community politics by attending Loretto town meetings to present their project to the town mayor and borough council. After collecting radon data in the community and on campus, the club invited all stakeholders to a poster session where an active data discussion ensued. That same year, the students handed out free radon testing kits at a local festival—providing over 200 kits to the community in less than two hours.

Conference Travel

When students engage in undergraduate research or community activities, the SFU chemistry faculty encourages the students to present at local, regional, or national events. The costs for local events can often be handled within the campus club budgetary limit as funds are provided by the SFU Student Government Association. However, travel to regional or national conferences often requires fund-raising or students have to pay out of pocket. Students are encouraged to apply for travel funding and a number of opportunities are available on campus, including university travel grants through the School of Sciences or the SFU Office of Undergraduate Research grant program.

Additionally, many local ACS sections, such as the Pittsburgh Section, have travel awards for undergraduate researchers (*8*). Students have the opportunity to apply twice each year (once for each national meeting) and, if funded, are required to write a report to the section, which is published in the local newsletter.

Sources of Funding

Funding sources mentioned above are organized below to clearly delineate the variety of resources.

Campus Funding

Becoming a recognized organization on campus opens up many doors to funding. The Saint Francis University Student Activities Organization (SAO) funds all recognized student organizations at some financial level. However, to be recognized, the SFU Chemistry Club must have a constitution, elect officers annually, and follow the rules, including having at least one fund-raiser each year. The Club must also provide a report each year on expenditures and submit and defend a proposed new budget each March. The SFU SAO has been a steady source of funding from ~$400–1,000 per year depending on the ability and ambitions of the officers.

Once the SFU Chemistry Club is recognized on campus additional funding doors on campus open, including requesting money from the Chemistry Department, the SFU School of Sciences, the SFU Provost's office, and the SFU President's Office. Each administrative unit has a small discretionary fund that may be tapped if the conditions are right (right idea, right time, money available, etc.). Fund-raising from administrative units often depends on the quality and interests of the officers.

Corporations

Another funding avenue is to look for opportunities with local companies and corporations and take the time to build relationships. Since the SFU Chemistry Club is so active in community service, the club often needs transportation to and from off-campus events. The cost of using a campus vehicle became prohibitive (as the campus charged for mileage). However, one member of the Board of Trustees owned a number of auto dealerships and also was a chemical engineer by training. By working with the SFU Advancement office and building a relationship with the board member, a chemistry (ROCK) van has been donated every seven years for outreach events—with their company name prominently shown on the vehicle!

Government Programs

State government programs also provide an opportunity for chemistry clubs. In Pennsylvania, the state runs the Educational Improvement Tax Credit Program (EITC) (*9*). An organization (university) has to apply to be an Educational Improvement Organization (EIO) and collect letters of support from public school district superintendents. Once approved by the state, the organization (SFU) is listed on a state website and companies can donate money to the SFU outreach program in exchange for tax credits. Through the EITC program, the SFU Chemistry Club has been supported by banks and other private corporations. The state requires that the money donated to the program only be spent at the public schools from which grant recipients have letters and each year the recipient organization must file a report detailing the money spent, any amount carried over, and a list of events/activities completed using the money.

Foundations

Working with your university Grants and Foundation office will help to find additional resources for funding. It is important to have an idea in mind or a project you want to complete. When you approach the grants officer be prepared to explain your project and they can help you find foundations to fund projects in your interest area. In addition, if the grants office knows what you do, they will seek you out, as the officers want to/need to bring in external funding for the university. With the help of the SFU Grants and Foundation office, we have been able to develop projects funded by Highmark Blue Cross Blue Shield, the Conemaugh Health System, and the Buhl Foundation for SFU Chemistry Club outreach efforts.

Additionally, the SFU Chemistry Club outreach efforts have been funded by the Pennsylvania and West Virginia Campus Compact to assess the impact our events have on K–12 students and SFU students (*10*).

103

American Chemical Society

Of course, the American Chemical Society (ACS) has a number of programs to support chemistry clubs across the nation. Current programs include (but are not limited to): National Meeting Travel Grants, Local Section Travel Grants, Community Interaction Grants, and New Activities Grants (5). Working with the local ACS section may also lead to funds.

Chemistry Honor Societies

Chemistry honor societies such as Gamma Sigma Epsilon (GSE) or Phi Lambda Upsilon (PLU) often have memberships that overlap with the chemistry club and may provide opportunities for both clubs to succeed (11, 12). By coordinating activities, both organizations can fund-raise a little and maximize the impact.

Non-ACS Scientific Societies

Other local scientific societies are willing to fund chemistry club/outreach activities. In SFU's region, we are fortunate to have the Society for Analytical Chemists of Pittsburgh (SACP) and the Spectroscopy Society of Pittsburgh (SSP). Both are non-profit scientific societies devoted to the education and support of future scientists in the Pittsburgh region.

National organizations such as Sigma Xi fund research projects, not outreach or other chemistry club activities, but you can use this to your advantage by funding one of the community-based projects mentioned above.

Untapped Resources

Departmental alumni know your institution and understand your passion and your contributions to the community. In cooperation with your campus Alumni office (often combined with the Advancement office), you can reach out to your alumni base. In addition, you can reduce the chemistry club and faculty adviser burden by persuading the Alumni office of the benefits of fund-raising with the chemistry club. It is important to have the right project at the right time.

Conclusion

While fund-raising is often seen as a chore or a challenge, it is a critical thinking endeavor and an important opportunity to help your chemistry club members gain valuable experience. The important skill of writing out and explaining an idea to a potential funding partner is invaluable. The skill of crafting an idea into a fundable project through revisions and refining the activity will serve your graduates in their future careers. Different campus offices (Provost, President, Advancement team, Alumni) can be partners in helping you meet your goals. Campus resources, such as the Grants and Foundation Office, are beneficial to help you find the right funding fit. Do not forget to search the ACS website for the targeted chemistry club grant opportunities. Finally, be willing to try some different fund-raising ideas—from selling donuts to having a bowling night—find what works for and excites your group of students!

References

1. Howard, R. J.; Ropp, J. A.; Wasil, C. M.; Zovinka, E. P. Rural Outreach Chemistry for Kids (R.O.C.K.) A Service Project to Involve More than Chemistry Majors in a Chemistry Club. *The Chemical Educator* **1997**, *2*, S1430–4171.
2. Minor, J.; Rosmus, T.; Zovinka, E. P. Science ROCKs at Saint Francis University…and Will Soon ROLL! *InChemistry* **2012** (April/May), 19–23.
3. Clark, R. A.; Fry, C. M.; Mosier, D. R.; Zovinka, E. P. Using Community Service Activities to Invigorate the SFU Chemistry Club. In *Building and Maintaining Award-Winning ACS Student Members*; Mio, M. J., Benvenuto, M., Eds.; ACS Symposium Series 1230; American Chemical Society: Washington, DC, 2016; Ch. 9, pp 95–108; DOI: 10.1021/bk-2016-1230.ch009.
4. Saint Francis University. www.francis.edu/franciscan-higher-education (accessed Nov. 25, 2017).
5. ACS Student Chapter Grants. www.acs.org/content/acs/en/funding-and-awards/grants/acscommunity/studentaffiliatechaptergrants.html (accessed Nov. 25, 2017).
6. Gloria Gates Memorial Foundation. https://ggmf.org/ (accessed Dec. 10, 2017).
7. Solomon, D.; Bopp, J.; O'Donnell, L.; Petrovic, J.; Snavely, R.; Zovinka, E. P. Using Campus Radon Testing as a Freshman Colloquium. *The Chemical Educator* **2003**, *8*, 37–40.
8. Pittsburgh Section of the ACS. www.pittsburghacs.org/awards/pittsburgh-section-travel-grants (accessed Dec. 8, 2017).
9. Pennsylvania Department of Community and Economic Development. https://dced.pa.gov/programs/educational-improvement-tax-credit-program-eitc/ (accessed Dec 6, 2017).
10. Lynch, M. T.; Zovinka, E. P.; Zhang, L.; Hruska, J. L.; Lee, A. M. Rural Outreach Chemistry for Kids (R.O.C.K.): The Program and Its Evaluation. *Journal of Higher Education Outreach and Engagement* **2005**, *10*, 125–141.
11. Gamma Sigma Epsilon. www.gammasigmaepsilon.org (accessed Nov. 25, 2017).
12. Phi Lambda Upsilon Grants. philambdaupsilon.org/grants/chapter-grants (accessed Nov. 25, 2017).

Chapter 9

ACS Student Chapter at the University of Northern Colorado: The Legacy of Professor Kimberly A. O. Pacheco

Murielle A. Watzky*

Department of Chemistry and Biochemistry, University of Northern Colorado, Greeley, Colorado 80639, United States
*E-mail: murielle.watzky@unco.edu

The ACS Student Chapter at the University of Northern Colorado mourns the loss of its long-time faculty mentor, Dr. Kimberly A. O. Pacheco, and reflects on the impact of her strong dedication, vision for excellence, and many lasting contributions. Nearly 57 years after its chartering, the Student Chapter actively participates in outreach activities around northern Colorado and coordinates social events for the Chemistry and Biochemistry community on campus. The structure of the Student Chapter is described herein, along with activities organized by Student Chapter members. The ACS Student Chapter recently received a Salute to Excellence award from the ACS Colorado Local Section. Dr. Pacheco's successful efforts at reaching out to other active ACS Student Chapters around Colorado are also discussed.

Kimberly A. O. Pacheco: Faculty Mentor for the ACS Student Chapter (2001–2016)

Many of us are lucky to have crossed paths with mentors who have had a strong positive impact on our lives. Dr. Kimberly A.O. Pacheco (Figure 1), who was the faculty mentor for the ACS Student Chapter at the University of Northern Colorado (UNC) from 2001 until her death in 2016, not only built an award-winning chapter but also generously shared her vision and leadership with many young minds and future chemistry professionals, and she did so unconditionally.

Figure 1. Dr. Kimberly A. O. Pacheco

Dr. Pacheco joined the Department of Chemistry and Biochemistry in 2001, and was a Professor of Organic Chemistry when she passed away after a 9-year battle with cancer. She was a role model to her students, both undergraduate and graduate, and was an inspiration to junior faculty. Even while fighting cancer, she would find the strength to provide guidance and genuine support. I am lucky to have known her during her last few years at the University of Northern Colorado. As with her students, she always expressed confidence in me. Professionally we were, after all, just "diamonds in the rough".

ACS Student Chapter at UNC

The ACS Student Chapter at UNC, a public comprehensive university located in Greeley, Colorado, was chartered on June 12, 1961, under the institution's former name of Colorado State College. UNC serves a diverse undergraduate population that includes over 35% of first-generation students.

Recent records show that the ACS Student Chapter received several Honorable Mention Awards between 1997 and 2000. Under Dr. Pacheco's mentorship from 2001 to 2016, it also received Commendable and Outstanding Awards (Table 1).

More than 55 years after its chartering, the ACS Student Chapter is still actively contributing to the community at large. Each year, Student Chapter members carry out multiple outreach activities at local fairs and K–12 schools, and coordinate events for the student community on campus and for the Department of Chemistry and Biochemistry. Student Chapter members have also contributed to the Undergraduate Programming at Regional ACS Meetings. The ACS Student Chapter has recently been recognized with several awards from the community and is sporting a new design (Figure 2).

Table 1. ACS Student Chapter Awards at UNC Under Dr. Pacheco's Mentorship

Academic Year	ACS Student Chapter Award(s)
2012–2013	Commendable Award
2011–2012	Honorable Mention Award
2008–2009	Commendable Award
2007–2008	Outstanding Award/Green Chapter Award
2006–2007	Commendable Award
2005–2006	Outstanding Award
2004–2005	Commendable Award

Figure 2. ACS Student Chapter design at UNC (Syd Dolzine).

Student Chapter Members: A Community of Chemistry Ambassadors

The Department of Chemistry and Biochemistry at UNC usually encompasses over 150 students registered as majors across all 4 years and graduates a class of fewer than 30 seniors each spring. Excellence in teaching established by longstanding faculty members, along with a low student-to-faculty ratio, have helped build a convivial and collegial community among majors.

In their roles as ACS Student Chapter members, Dr. Pacheco wanted her students to have opportunities to value the field of chemistry while growing as chemistry ambassadors to the community at large.

Student Chapter Members

New Student Chapter members are recruited every year with the help of advertising pamphlets, which are distributed at the department's fall BBQ cookout. Throughout the academic year, members of the ACS Student Chapter hold biweekly "pizza night" meetings, during which all outreach, fund-raising, and social activities get planned and volunteers are recruited. The Student Chapter maintains visibility with a web page accessible from the department website and holds a presence on social media platforms.

Student Chapter Officers

Student Chapter officers are elected each year by Student Chapter members and get introduced at the department's spring Chemistry Awards banquet. Student Chapter officer positions are described in Table 2. In 2018, Student Chapter officers received a Salute to Excellence Award from the ACS Colorado Local Section for their work with the Undergraduate Programming at the 2017 ACS Rocky Mountain Regional Meeting. In 2017, the Student Chapter President received an Undergraduate Student Leadership Award from the College of Natural and Health Sciences at UNC.

Student Chapter Activities: Serving the Department, Campus, and Northern Colorado Communities

Dr. Pacheco felt that an active ACS Student Chapter should provide its members with a sense of community, opportunities for outreach, a healthy competitive spirit with other clubs while gaining a higher profile on campus, and connections (*1, 2*).

Sense of Community

While reinforced by all Student Chapter activities (Table 3), a sense of community can start with social events aimed at Student Chapter members. The Student Chapter holds biweekly meetings with free pizza, facilitates study nights, and coordinates group hikes. The Student Chapter also organizes events aimed at the broader community of the Department of Chemistry and Biochemistry, both faculty and students, such as the popular fall BBQ cookout and the formal spring Chemistry Awards banquet.

Table 2. ACS Student Chapter Officers

Officer Position	Officer Role(s)
President	• Supervise student organization meetings, events planning, and list of members • Submit annual report of student organization activities and required paperwork for student organization renewal • Take on role of other officers, if needed
Vice President	• Assist president in student organization meetings and events planning • Schedule fund-raising events • Coordinate budget and fund-raising events with treasurer
Treasurer	• Create budget for Student Chapter • Make purchases on behalf of Student Chapter • Balance Student Chapter bank account • Generate fund-raising events
Secretary	• Take notes at meetings • Collect photos at events • Keep a history of events throughout the year
Public Relations Officer	• Maintain Student Chapter web page and social media • Coordinate notes and photos with secretary
Demonstrations Officer	• Organize and perform chemistry outreach demonstrations • Enforce safety requirements

Opportunities for Outreach

The Student Chapter participates in a number of outreach events throughout the year (Table 3). Some may be university-sponsored and on campus (UNC Community Fest and Mathematics and Science Teaching Institute at UNC), away from campus (Denver Maker Faire) or within the K–12 community in northern Colorado. A recent demonstration at a local elementary school generated a heart-warming outcry from a fifth grader who exclaimed, as our all-female cast of undergraduate students walked in with goggles, "I didn't know scientists could be girls!!!"

Higher Profile on Campus

Student Chapter members get to work side by side with other clubs during university-sponsored events, such as the UNC Community Fest and at the Denver Maker Faire, and participate with other clubs in campus events, such as the Relay for Life. Our Student Chapter also gains visibility with Mole Day contests and a Periodic Table cupcakes sale (Table 3).

Table 3. ACS Student Chapter Annual Activities

Activities serving Student Chapter members	• Biweekly pizza night meetings • Study night sessions • Group hikes in nearby Rocky Mountains
Activities serving the Department of Chemistry and Biochemistry	• Fall BBQ event 　◦ Cookout 　◦ Yard games 　◦ Liquid N_2 ice cream • Spring Chemistry Awards Banquet 　◦ Outstanding Student Performance Awards 　◦ Distinguished Alumni Award 　◦ Graduating Seniors recognition
Activities serving UNC students	• Mole Day contests • Periodic Table cupcakes • ACS Student Chapter t-shirt sale • Goggles sale • ACS Study Guides sale
Activities serving the northern Colorado Community	• UNC Community Fest 　◦ Chemistry hands-on demonstrations 　◦ One-day event serving hundreds of attendees • Denver Maker Faire 　◦ Chemistry hands-on demonstrations 　◦ Two-day event serving thousands of attendees • Chemistry outreach demonstrations at local K-12 schools • Mathematics and Science Teaching Institute at UNC 　◦ Chemistry outreach demonstrations for local K–12 students • Longs Peak Science and Engineering Fair prizes for the "Schrödinger's Outside the Box Award for Creativity and Innovation" • Relay for Life team (American Cancer Society) • Food drive participant (Weld County Food Bank)

Connections

Dr. Pacheco initiated the participation of Student Chapter members in events organized by the ACS Colorado Local Section, thus providing students with the opportunity to form connections both outside the department and outside the university (see next section).

A Community of Local ACS Student Chapters

Colorado Student Affiliates Conference

Dr. Pacheco co-organized Colorado Student Affiliates Conferences (COSACs) with Dr. Sandra Bonetti of Colorado State University Pueblo in 2007 and 2008. The conferences met over a full day and were established "with the purpose of strengthening, activating, and revitalizing existing affiliates, and assisting with start-up activities for new chapters" (*1*).

The conferences consisted of "an interactive forum with informative presentations, workshops/discussions led by facilitators, [and] highlight(s) of student affiliate groups' activities with demonstrations and a poster session" (1). Topics of interest included How do you start and revive an affiliate group?, What types of activities do affiliate groups sponsor and perform?, and What role do student affiliates play in the national ACS and the ACS local section (2)? Examples of tasks assigned for discussion at COSAC are shown in Table 4 (2).

Table 4. Tasks Assigned for Discussion at COSAC

What has prevented the chapter from being more active than it is currently? How can this problem be corrected?
What problems do you encounter when trying to organize events? How can these problems be corrected?
What are your most successful events at the campus level? In the community?
How could you collaborate with another ACS Student Affiliate Chapter for an event? (National Chemistry Week?) Identify chapters to contact.
How could the ACS Colorado Local Section help facilitate an event for your chapter?

National Chemistry Week

Dr. Pacheco served as the National Chemistry Week Chair for the ACS Colorado Local Section, starting in 2002 (3). From 2004 to 2011, she organized an annual outreach event with hands-on activities for elementary school students, in collaboration with the Denver Museum of Nature and Science. The event took place at the museum and was run by members of local ACS Student Chapters. In 2008, a crowd of 6,513 attendees was reportedly present at the museum during the event.

Undergraduate Programming at Regional ACS Meetings

Dr. Pacheco coordinated the Undergraduate Programming at the 2012 ACS Rocky Mountain Regional Meeting, along with General Chair Dr. Connie Gabel from Metropolitan State University of Denver. The program included poster sessions, a "Careers in Chemistry" panel, and social events, such as a scavenger hunt and marshmallow roast (4). These activities served as the blueprint for the Undergraduate Programming at the 2017 ACS Rocky Mountain Regional Meeting (5).

Councilor for the ACS Colorado Local Section

Dr. Pacheco served as Councilor for the ACS Colorado Local Section from 2003 to 2011 and received the ACS Colorado Service Award in 2010 for "outstanding and tireless efforts promoting our profession". Dr. Pacheco's work at the Denver Museum of Nature and Science helped contribute to a 2009

ChemLuminary Award to the ACS Colorado Local Section for Outstanding Performance.

Acknowledgments

My warmest thanks to members of the ACS Colorado Local Section executive committee for providing necessary information: Drs. Sandra Bonetti, Connie Gabel, Margaret Rakowsky, and Susan Schleble. Special thanks to Charles Pacheco and Dr. Jetty Duffy-Matzner for their help. Additional thanks to Dr. Pacheco's former graduate students and to current UNC ACS Student Chapter officers for their support.

A Chemistry/Biochemistry Undergraduate Research Endowment under Dr. Kimberly A. O. Pacheco's name is available with the UNC Foundation (uncf.information@unco.edu).

References

1. Pacheco, K. A. O. Starting/Reviving and Sustaining a Thriving Student Affiliate Chapter. Presented at Colorado ACS Student Affiliates Conference, Metropolitan State University of Denver, October 2007.
2. Pacheco, K. A. O.; Bonetti, S. J. Starting/Reviving and Sustaining a Thriving Student Affiliate Chapter. Presented at Colorado ACS Student Affiliates Conference, University of Colorado at Colorado Springs, October 2008.
3. ACS National Chemistry Week Home Page. American Chemical Society: Washington, DC, 2018. https://www.acs.org/content/acs/en/education/outreach/ncw.html?cid=ad_onlinestore_shippingemail_ncw_sep-oct_2016 (accessed April 29, 2018).
4. 2012 Rocky Mountain Regional Meeting Home Page. American Chemical Society: Washington, DC, 2012. http://rmrm2012.sites.acs.org (accessed April 29, 2018).
5. 2017 Rocky Mountain Regional Meeting Home Page. American Chemical Society: Washington, DC, 2016. http://rmrm2017.sites.acs.org (accessed April 29, 2018).

Chapter 10

Innovation for a Healthier Student Chapter: Student Activities at the University of New England

Molly V. Wright and Amy E. Keirstead*

Department of Chemistry and Physics, University of New England, 11 Hills Beach Road, Biddeford, Maine 04005, United States
*E-mail: akeirstead@une.edu

The University of New England American Chemical Society Student Chapter is a recently reactivated chapter that employs chemistry as a vehicle for community outreach, environmental and global improvement, and camaraderie, stemming from a shared interest in the sciences. Here we draw upon specific activities, methods, and actions perfomed by the club, detailing its evolution from reactivation to receiving the Outstanding and Green Chemistry Awards. The focus of this text is to introduce and examine potential routes for success experienced by this chapter, which may be of particular interest for similarly (re)activated chapters.

Introduction

The motto for the University of New England (UNE)—a comprehensive university with three campuses located in Biddeford, Maine; Portland, Maine; and Tangier, Morocco—is *Innovation for a Healthier Planet*. This motto can be applied in myriad ways to our diverse program offerings, including our graduate degrees in the health sciences, our interdisciplinary research in the marine sciences, or even the environmental awareness theme that runs through our undergraduate core curriculum. Our UNE Chemistry Club is also an embodiment of the innovation for which we strive at UNE and the health that follows as a result. By striving for continual improvement and reaping the benefits, we follow the path of progression that is essential for a healthy ACS student chapter. Innovation is a process, and so is the development of our club. In this chapter, we will guide you through the evolution of our ACS student chapter—where we came from, how we gained momentum, and where we are going.

Where We Have Come From: Reactivation and Revival (2013–2015)

With graduate programs in osteopathic medicine, dental medicine, and pharmacy, as well as popular programs in marine science, neuroscience, and animal behavior, many of the undergraduates at UNE are in science-based majors and take chemistry courses; however, the number of majors in chemistry and biochemistry is comparatively small. Still, UNE students have a long history of active participation in the UNE Chemistry Club, which had previously been an ACS Student Affiliate group, but over the years regressed to "inactive" status. The membership in previous iterations of the club primarily consisted of chemistry and biochemistry majors with an emphasis on K-12 outreach in local schools, including classroom visits and science nights. The club was also a useful vehicle for developing strong camaraderie among classmates and relationships with faculty.

Following some faculty turnover and new faculty advisor leadership, our club began the process to reactivate as an ACS student chapter in the fall of 2013. The reasons for pursuing reactivation were numerous: First, our club members were increasingly interested in professional development. Reactivating as an official student chapter would provide access to a wide array of ACS resources including graduate school, career preparation, and webinars. Second, our club members wanted to diversify their outreach opportunities beyond K-12 work and traditional "demos," and closer ties to ACS would result in more guided outreach ventures including National Chemistry Week (NCW). Third, faculty in our department had been sending students to present their undergraduate research at ACS national meetings for a number of years, and the recognition as an active chapter would afford us the opportunity to apply for National Meeting Travel Grants and other funding opportunities. Finally, our club was seeking revival and renewed direction, and participation in the ACS Student Chapters program would mean annual self-reflections and reports, as well as reviews from a panel of faculty offering new ideas and suggestions for how to improve. The club was able to meet the requirement of a minimum of six ACS student members through cooperation with our club's host department, the UNE Department of Chemistry and Physics (DCP). The DCP offered to pay the membership fees for its majors who were in their junior year, as a resource for the students' professional development, in exchange for the students' help with tasks such as open houses and hosting visitors. The DCP has continued to provide this opportunity in subsequent years; however, some students who are in their freshman, sophomore, and senior years choose to become ACS members at their own expense to take advantage of the periodic table blankets offered by the Member-Get-A-Member program. (When the DCP covers the cost of the student membership, the blankets are given to our Chemistry Club for use as door prizes, raffles, etc.)

Chapter membership in the years following reactivation (2013–15) was primarily comprised of chemistry and biochemistry juniors and seniors with the traditional officer roles of president, vice president, secretary, and treasurer. Recruiting efforts in these early years were minimal, with word-of-mouth being the primary mechanism and a focus on chemistry and biochemistry majors. Undoubtedly the limited membership was linked to these limited efforts! Indeed, "elections" for officer positions (held in September) were usually by acclamation, or the process of the faculty advisor convincing a member with club history, leadership abilities, and/or organizational skills to take on an officer role. Three faculty advisors supported (and continue to support) our chapter. The primary advisor is a chemistry faculty member who assists with any administrative needs (including liaising with ACS and our Student Clubs and Organizations Office), financial oversight, and general advising/coordination. Two co-advisors (a chemistry staff member and a physics faculty member) support our club by assisting with our outreach activities and event planning and fund-raising, respectively. We implemented this model initially because our co-advisors wanted to support our club but had limited expertise in chemistry or held a position at the university that precluded them from taking the "faculty advisor" title; additionally, we needed an ACS member as a faculty advisor. That being said, this tripartite model works well for us, with one faculty advisor overseeing the finances and general coordination/liasing and two other department members supporting the students in their own niche areas.

Funding for club activities in those early years was provided by the UNE Undergraduate Student Government. All students pay an activity fee, a portion of which is distributed to student clubs, which submit a budget request each spring, and in turn, the club president and treasurer are required to attend a training session in the fall.

Our goals as a newly reactivated chapter were fairly simple: We wanted to provide more opportunities to learn about chemistry for both our members and the general student body, and we wanted to diversify our outreach ventures by celebrating NCW, which had not been celebrated on campus in at least five years. In these first two years, our club partnered with the UNE Math Club for field trips to Boston, (about 90 min away) as well as visits to the science museum, the aquarium, and a local nanotechnology company for an information session about quantum dots and a tour. These field trips were fantastic for generating collaboration and camaraderie with another club and served as a way to share our limited resources while providing new opportunities and information to our members. Our club also partnered with UNitEd, the multicultural club on campus, to host a showing of *Forgotten Genius* during Black History Month, followed by a facilitated discussion. This was an excellent way to explore some of the history of chemistry and better understand the challenges faced by African-American chemists.

Social opportunities and generating a sense of camaraderie have always been important to our students and in these early years the club hosted a biannual "Trivia Lunch" where students and faculty formed teams to answer skill-testing questions about chemistry, general science/general knowledge, and "fun facts" about the professors. This event quickly became a favorite for UNE students, in part because a pizza lunch was provided by the chemistry club, and also because it served as an opportunity for students and faculty to socialize and collaborate outside of the classroom setting. Additionally, it allowed for the development of a friendly rivalry between teams competing to become the next trivia "champIONS." Another annual tradition—our holiday party, where we make ornaments out of chemical glassware—was born during our first year of reactivation and became an instant hit. The opportunity to relax while eating cookies and drinking hot cocoa fostered a creative outlet. Making snowmen or sparkly ornaments out of unused but unusable glassware (boxes of which were unearthed during a lab renovation) became a favorite for our students, many of whom would bring friends to share in the fun. In fact, our current chapter president got involved with the club after being invited to this party during her sophomore year!

In those early years, we also continued our involvement in local K-12 outreach, much of which involved on-campus visits from local school groups through a National Science Foundation-sponsored program in which masters of science students in the marine science department volunteered in local schools; one of our co-advisors was instrumental in liaising with this group and coordinating these visits. However, celebration of NCW remained a key goal, and thus "Chemapalooza" was born—an annual one-day celebration of chemistry using that year's NCW theme. The events were held during lunchtime in the function rooms adjacent to the cafeteria to attract maximum traffic, and faculty in the DCP supported the events by promoting them within the classroom setting and attending events when possible. The festivals included traditional standbys such as chemistry trivia and liquid nitrogen ice cream, and the themes were incorporated into the event through hands-on activities and games such as "Buckyball bowling" for the nanotechnology theme or "Mentos in Coke" for the chemistry of candy theme. In retrospect, our Chemapalooza activities were more entertainment than education, and we could have done a much more thorough job of explaining the chemistry behind the fun. But the events did serve two important purposes: They gave our chapter an opportunity to plan and execute a recurring large-scale event on campus for our university community, and they increased awareness of our club on campus. People began to recognize Chemapalooza as our "signature event" and would ask what we were planning for the coming year.

Although our chapter made significant progress in those first two years after reactivation, and we were recognized with two successive Honorable Mentions through the ACS Student Chapters Awards program, we experienced some growing pains and were not without challenges. First, we struggled with our membership numbers; while we had some amazing and dedicated members, as well as devoted faculty advisors, both recruitment and retention of members remained a challenge. Admittedly, our recruitment efforts were poorly timed and uncreative, re lying mostly on word-of-mouth. We held our elections in the fall of each year, which made it difficult to maintain momentum when we had to start from scratch the following year, and the new officers did not have the existing officers to help guide them in their new roles. One significant challenge arose in our first year after reactivation, when our vice president left the university a few weeks into the fall semester and our president did not return for the spring semester. With these losses came new opportunities: a student who had recently transferred to the university, but had showed dedication, enthusiasm, and leadership qualities, stepped up and took on the role of president for the remainder of that year and the following year. This added continuity helped to balance out the juggling act of the previous year.

Our goals of "starting small" and focusing on one key outreach event and one key professional development event per semester, along with three spaced-out "signature" social events per year, allowed us to balance some of these growing pains and really work out the kinks of organizing and executing these new activities. The process of keeping track of events using the Student Chapters Online reporting system, as well as having to do a year-end self-evaluation and set goals for the coming year, was new to us but proved to be beneficial in terms of gaining some traction. The feedback and suggestions provided to us by the faculty reviewers through the ACS Student Chapter Awards program were invaluable for our growth, and we began to get more invested in the ACS Student Chapters program and more aware of the resources available to us. Earning the "Honorable Mention" awards was an achievement we were proud of, but we knew that we could do more and achieve greater things by learning from our growth and development process, brainstorming innovative and creative ideas that would interest a wide array of students, and taking advantage of the many opportunities provided by ACS as well as other organizations.

Gaining Momentum: Growing and Evolving as an ACS Student Chapter (2015–2017)

By the fall of 2015 our UNE Chemistry Club had made enormous progress, going from an inactive ACS student chapter to one worthy of Honorable Mention in just two short years. Not only were we being recognized by ACS itself but by the university community as well. Certain signature events such as Chemapalooza (NCW), biannual trivia lunches, and the holiday ornament decorating party started to be anticipated by students across the disciplines, creating a campus atmosphere that was receptive to new events and activities hosted by the club. This new recognition could be attributed to several changes made in order to improve upon and boost our initial growth such as restructuring elections, revamping our recruiting efforts, and, consequently, diversifying membership.

We restructured our elections by holding them near the end of the spring semester rather than in the fall, which resulted in a much more efficient turnover and allowed current club officers to mentor their successors. This informal training process provided incoming club officers of the opportunity to study the role they would be filling and learn how to best meet and exceed the expectations of that role. Our club began holding "retreats" at the end of each school year wherein the newly elected officers would meet with their predecessors to both wrap up the current year and get a jump start on the year to come. During these retreats, the group would engage in a roundtable review and assessment of the year at hand, discussing which events worked well and which ones did not, and why. Following lunch, the outgoing officers would complete the

Student Chapters Online report, while the incoming officers would brainstorm and compile a calendar of activities for the coming year. This team-building exercise and planning for the year ahead prevented stagnation within our club that had previously taken place over the summer between officers, and the organized reflective and goal-setting processes were instrumental in helping our club continue its growth and progression. Recognizing the demands of keeping up with both our social media platforms and the ACS Student Chapters Online reporting system, we rewrote our Club Constitution to include the position of club historian, who would fulfill these roles as well as maintain an online repository of photos.

As our ambitions as a student chapter grew, so too did our need for increased membership to enable us to execute all of our planned activities. Thus, we began to invest more time and effort in member recruitment and retention. By utilizing university resources such as the GetInvolved portal (a web-based club networking system) and setting up a display at the Club Involvement Fair, not only could a greater number of members be attracted, but a greater variety of students could be reached. Knowing that we faced the challenge of having to overcome the (often) negative stigma of chemistry, we worked to make our recruiting efforts fun, giving away Nerds candy in a test tube adorned with the details of our first meeting, or having as a door prize a periodic table blanket (from the ACS Member-Get-A-Member program) at the first meeting that could only be won by someone in attendance. Implementing feedback from one of our chapter reviews, we started holding social events early in the fall semester to make new members feel welcome and give everyone a chance to reconnect. Our club also started using the social media platforms Facebook and Twitter to advertise upcoming events and promote fun activities, as well as a means to share resources from ACS. Finally, the club designed a logo using element letters to spell out our name, which we incorporated into posters and other advertising, allowing us to start "branding" ourselves. We also purchased T-shirts with our logo and a chemistry joke on the back and asked members to wear them at events to further promote our club and look more like a professional team. Overall, membership recruitment became less centered solely around chemistry and biochemistry majors, allowing for both greater numbers and greater diversity, which were essential for bringing new ideas to the table. While we did not retain all of these new members, our numbers and involvement grew meaningfully.

With increased membership and greater circulation of new ideas, we began to have the organization and the "people power" to improve upon our signature events and incorporate new ones as well. National Chemistry Week became our chapter's biggest celebration, morphing from one day into a week-long event to celebrate the various yearly themes and promote both our club and chemistry in general. By reserving an outdoor space in the campus quad for the lunch hour (rain location: inside, near the cafeteria), we had significantly more foot traffic with students, staff, and faculty alike stopping by our table to find out what we were up to that day. We have used this format for the last three years, including popular standbys such as a periodic table of cupcakes (Monday) and liquid nitrogen ice cream (Friday), along with daily chemistry-themed door prizes. Hands-on activities and demonstrations incorporating the year's theme round out the week and allow us to touch upon more of the science behind the fun that had been missing in previous years. The biggest struggle in bringing Chemapalooza to fruition has been planning and preparation, but over the years we have developed a system that works for us. While the officers take the lead in planning for the week, it is really an "all hands on deck" approach, and we use online tools for planning and scheduling so that everyone has access to all materials and a thread of e-mails is not necessary. Club members are asked to sign up for varying time slots two weeks in advance to ensure that there are people to run the activities throughout the week, and the members running the events practice and finalize the demos at the meeting prior to NCW. We have used the Sunday before NCW as a chapter-building event where we go apple picking, followed by lunch and cupcake decorating at the faculty advisor's home. Getting together the day before allows us to feel more like a team and also gives us the chance to iron out any final wrinkles. Although last-minute problems always arise, these simple steps have allowed us to follow through with our big plans.

As signature club events began to grow, new resources were being sought and utilized as well, particularly in the area of professional development. One key resource for our student chapter has been the ACS Program in a Box. We had long talked about showing ACS webinars as professional development, but the catalyst for implementation was the introduction of these student-friendly webinars with a box of door prizes and activities; all we needed to do was book the room and bring some snacks! Interdisciplinary topic webinars were incorporated in order to include other clubs (such as exercise science for *The Chemistry of Sports*) and, while attendance is often small, the benefits of these webinars for our members are significant. Another key professional development activity has been our chapter's more focused participation at ACS national meetings. While students from our club had attended national meetings previously, we only began attending as a chapter in the fall of 2015, when we sent several members to nearby Boston to present undergraduate research and participate in the undergraduate program including presenting a poster at the Successful Student Chapters (SSC) session. Three members traveled to San Francisco in the spring of 2017 to present their own research as well as an SSC poster and to participate in the Chem Demo Exchange. The myriad events and programs provided for undergraduates and future chemistry professionals at these ACS meetings afford invaluable professional development for our members, and planning for and traveling to the meeting is an important chapter-building exercise. Finally, we were fortunate to be involved in a local section meeting held at our institution and hosted by our faculty advisor in spring of 2017 at which Dr. John Warner (one of the founders of green chemistry) gave a guest lecture. Our chapter set up two tables to promote and educate visitors about green chemistry, attended the dinner where local section business was discussed, and formally thanked Dr. Warner after his talk.

Another key resource for us has been our collaboration with Beyond Benign, the green chemistry education nonprofit located in Massachusetts. In the fall of both 2015 and 2017, chapter members were invited for an on-site workshop to train as Green Chemistry Student Outreach Fellows, where we and students from other local colleges learned about green chemistry and how to spread the message by doing outreach activities with K-12 students. As part of the program, fellows are asked to complete a certain number of outreach activities throughout the year using activities designed by Beyond Benign and with materials supplied by them. Our ongoing collaboration with Beyond Benign has afforded a wide array of outreach and professional development opportunities for our members, including being invited to green chemistry talks at local universities and getting together with other members of the Fellows' group at ACS meetings. Our collaboration with Beyond Benign—and the resources and opportunities afforded therein—has been instrumental in helping our chapter earn the Green Chemistry Award for the past two years. By partnering with Beyond Benign and focusing on green chemistry for our outreach activities, along with other educational opportunities, the UNE student chapter is utilizing our own discipline as a means of striving for a healthier planet, congruent with our university's mission.

Participation in the Maine Science Festival (MSF) was yet another new activity taken on by our chapter in spring 2016 and continued in 2017. This relatively new statewide festival is held annually in March and brings together an assortment of educational, science-affiliated groups that offer demonstrations, hold workshops, and lead various sessions encompassing myriad topics in science. In both 2016 and 2017, our Green Chemistry Outreach Fellows used the venue for their outreach efforts, and having the ready-made activities and most of the materials supplied by Beyond Benign made planning and preparation straightforward. The format for our activities at the MSF has varied between a large expo-type setting and a more focused workshop approach depending on the activities, which have ranged from biomimicry and eco-friendly packaging materials to construction of dye-sensitized solar cells. Participating in the MSF was just one more way for the club to utilize our new resources and engage in some professional development while giving back to the community.

An additional initiative during this period of club evolution and growth included a greater focus on fund-raising. We are fortunate to obtain the majority of our funding from the university, but with our club taking on so many new projects all at once, it was important to acquire additional funds that could be used at our discretion without limitations set by the university. One fund-raiser that we have run for two consecutive years consists of making and selling chemistry-themed valentines during the second week of February. A meeting was dedicated to writing and assembling the valentines, which consisted of plastic, food-safe test tubes filled with Nerds candies and a small card attached inscribed with some form of cheesy, science-related "love pun." The valentines were sold for $1 outside the cafeteria during the week leading up to Valentine's Day. Although we do not raise significant funds from this endeavor, the crafting session is a valuable chapter-building exercise and people have a lot of fun with the valentines, which serve to promote our club and lessen the negative stigma of chemistry. This past year, we had a lot of leftover test tubes due to snow days (university closure) so we instead attached tags with our chapter's logo and links to our Facebook and Twitter pages and handed them out during Accepted Students' Day in the spring in an effort to recruit new members for the following year.

Our third and fourth years following reactivation were busy ones, characterized by rapid evolution and growth (both quantitatively and qualitatively), forging new relationships, and exploring new opportunities while maintaining important collaborations and traditions. Increased member recruitment, generation of new and innovative ideas, and utilization of available resources, coupled with teamwork and the strong leadership skills of our members, allowed us to set key goals and then make measurable progresstoward them. Our efforts were recognized with the Outstanding and Green Chemistry Awards during the 2015–2-16 and 2016–2017 academic years, as well as a Chapter Spotlight in the Nov/Dec 2015 issue of *inChemistry* magazine (*1*).

Where We Are Going: Looking Forward (2017–2018 and Beyond)

With a solid membership base, a reasonably established "routine" of events and activities, collaborations with other clubs and organizations, and an effective election and officer transition plan, the UNE Chemistry Club has found its footing and is poised to continue to make meaningful strides toward its goals of promoting chemistry (including green chemistry) in the university community, providing professional development opportunities, and generating a sense of camaraderie among its members.

Recognizing the importance of social media and "branding," the club designed a new logo in the summer of 2017 that was professionally created by a graphic designer. The logo illustrates the key elements of our chapter: affiliation with our university (a blue test tube holder with the UNE logo), a love for chemistry (test tubes framed in a hexagon), inclusion and diversity (rainbow colors), and green chemistry (leaf). We have used this logo on all of our promotional materials and branding since its design and it is beginning to become widely recognized around campus. Having an attractive and meaningful logo is also a source of pride for our chapter's members.

One of the suggestions often included in our past ACS student chapter reviews is to partner more with other local student chapters. Unfortunately, there are currently no other active student chapters in our local area and, given the size of the state of Maine, it is challenging to connect with other groups. We attended a student leaders' workshop hosted by our local section in Fall 2017, where we got to meet and network with other student chemistry clubs from across the state, some of which were just getting started. (A short report on the workshop and a photo of our chapter officers was included in *Chemical and Engineering News* (*2*).) At this event we were able to share our own experiences as a relatively young student chapter; we plan to continue this partnership by meeting up with these groups at events such as the ACS National Meeting and the Maine Science Festival. If we are able to establish strong connections with other student chapters, we would like to work together on an ACS Inter-Chapter Relations Grant to promote a collaborative project bringing student groups together and providing an opportunity to share our passion for chemistry with communities in different regions of the state.

ACS Student Communities offers a significant number of grant programs for student chapters, one of which is the New Activities Grant that provides matching funds for a new project embarked upon by an active chapter. Given our unique location in the Northeast, as well as our university's focus on sustainability and ongoing development of an "edible campus," we researched and submitted a grant proposal in spring 2017 to tap the maple trees on campus, produce maple syrup, and study the chemistry of maple sugaring. Our funded project, *Tapping into Sweet Chemistry*, was supported by the UNE Office of Sustainability, Facilities Management, and our campus food provider, and started with a trip to a local maple syrup operation to learn about the process. In the fall, we tagged the sugar maples on our campus and added informational signs about our project and, in preparation for tapping (usually in mid-late February, depending on weather), we spent some time learning more about the chemical composition of maple syrup and purchased the necessary materials. We held a "boil off" on campus, attended a pancake breakfast on Maine Maple Sunday, and will prepare a presentation (with samples!) for an upcoming ACS National Meeting. While we are still in the midst of this project, it has already reaped significant benefits for our chapter and its members. First, the process of grant-writing, undertaken by our current co-vice presidents, involved meeting with different parties on campus, researching the process, and creating a budget. Embarking on a large, innovative project like this—one that is different from what clubs on campus typically do—is exciting for our members and helps to draw attention to our chapter while showcasing real-world applications of chemistry. Of course, the financial support from ACS, along with support from our on-campus collaborators, has been invaluable for helping us start what we hope will be an annual tradition.

In 2014, UNE opened a study-abroad campus in Tangier, Morocco, outfitted with American-style science (physics, biology, and organic chemistry) laboratories. Having chapter members on our international campus yields new opportunities for our club to engage in outreach activities (including in the field of green chemistry) with an entirely new audience, truly living up to our university's motto of *Innovation for a Healthier Planet*. Of course, designing and executing K–12 outreach in an unfamiliar culture, community, and with unknown resources may prove to be challenging, but we are excited for the opportunity to branch out and try something innovative on an international scale.

While our UNE student chapter has grown and evolved significantly over the past five years, we still have areas that require improvement. Our recruiting efforts—including setting up at the Club Involvement Fair and visiting general and organic chemistry classes—have worked marvelously; in fall 2017, we had over 30 people attend our first two meetings! However, we are still struggling with retention of new members beyond the "core group," with the drop-off coming when the semester begins to pick up and when we really need help with NCW activities. Adopting a committee structure and starting the planning for NCW earlier should help with this, and we intend to implement some of the ideas listed in a recent *inChemistry* article (*3*). We also need to make more progress on our fundraising efforts; we have been quite successful in securing funding from our local section, department, and college to send students to ACS National Meetings, but we will need to raise some of our own capital to expand our attendance at these meetings in coming years. Fortunately, several ideas are documented in Volumes 1 (*4*) and 2 (*5*) of this series, and we welcome the challenge to implement these or our own innovative fundraising ventures.

Conclusion

In five short years, we have gone from a chapter of humble beginnings to one with impressive goals both met and in progress. The manifestation of these goals would not have been possible without the utilization of new ideas that our university so highly promotes as "innovation," as well as the significant time and effort dedicated to our chapter's future by the departments, organizations, and individuals that have supported us. Our goal in telling our story is not just to share the triumphs, but to bring to light the reality that progress is a process, and growth takes time. So, to the newly (re)activated chapters reading this book for wisdom—get started, implement new ideas, and join us in making a healthier planet. To create, to learn, to progress: These are the foundations of chemistry and the heart of our student chapter's success.

Acknowledgments

The foundation for a successful student chapter is rooted in the efforts and creativity of its members, and the same has been true for our chapter. While many students and faculty have contributed to our revival and evolution over the past five years, the efforts of Bronwen Boe, Megan Perry, Ryan Juneau, Jessica Woolf, Katie Chalmers, and Jessica White have been especially instrumental in our chapter's success. Chris Ambrose and Jill Tenny have provided invaluable support in their role as co-advisors, bringing unique expertise and ideas to the table. We are also grateful for the support fom the UNE Department of Chemistry and Physics, UNE College of Arts and Sciences, and the staff in ACS Student Communities (formerly the ACS Undergraduate Programs Office).

References

1. Lindsey, R. Chapter Spotlight—University of New England. *inChemistry* **2015**, *25*, 18.
2. Keirstead, A. Maine Local Section Convenes Student Chapter Leaders. *Chem. Eng. News* **2017**, *95*, 36.
3. El-Ashmawy, A. K.; Fultz, M. 9 Ways to Recruit Student Chapter Members—and Keep them. *inChemistry*, October 19, 2017. https://inchemistry.acs.org/content/inchemistry/en/student-chapters/recruit-members.html (accessed May 23, 2018).
4. *Building and Maintaining Award-Winning ACS Student Member Chapters Volume 1: Holistic Viewpoints*; Mio, M. J., Benvenuto, M. A., Eds.; ACS Symposium Series 1229; American Chemical Society: Washington, DC, 2016.
5. *Building and Maintaining Award-Winning ACS Student Member Chapters Volume 2: Specific Program Areas*; Mio, M. J., Benvenuto, M. A., Eds.; ACS Symposium Series 1230; American Chemical Society: Washington, DC, 2016.

Chapter 11

Community Service as the Cornerstone of the Xavier University of Louisiana ACS Student Chapter

Michael R. Adams,* Mehnaaz F. Ali, Candace M. Lawrence, and Janet A. Privett

Department of Chemistry, Xavier University of Louisiana, 1 Drexel Drive, New Orleans, Louisiana 70125, United States
*E-mail: mradams@xula.edu

Xavier University of Louisiana, with its unique identity as a Catholic institution that is also historically Black, has as its mission a goal of promoting a more just and humane society. The Xavier ACS Student Chapter is guided by this mission and has built much of its success on a strong program of community service. Outreach events in local schools at all K–12 levels are central to the overall program, but a wide variety of other service activities are included. The program has been sustainable through a development of dedicated student leaders and successful fund-raising, including ACS student chapter grants.

Introduction

The Xavier University of Louisiana Student Chapter of the American Chemical Society (ACS), known on campus as the ACS Chem Club, has a long record of providing service to the greater New Orleans area. Much of its work involves outreach activities in local schools and other STEM-related events for schoolchildren in the region, and these activities fit well with Xavier's mission to promote a more just and humane society. The organization's service program is the cornerstone of the overall success of the chapter and has led to the chapter receiving recognition from ACS as an Outstanding Chapter numerous times. The chapter also has been recognized twice in recent years as the Organization of the Year on Xavier's campus and has received special recognition from Xavier's Office of Student Services for having the most outstanding overall community service program among all student organizations. As others have noted, a strong program of outreach and community service can be a vital component in attracting new members and sustaining an active ACS student chapter (1). In a previously published broad overview of the Xavier chapter, it can clearly be seen that this, too, is the case for our chapter (2).

Overview

The successful outreach and service program has been built over the years through a number of specific steps. In the early 2000s, the leaders of the organization made the wise decision to create positions on the executive board for two community service co-chairs. These two students are tasked with developing community and campus service activities for the club and organizing members to successfully complete these events. Because this is their only assigned job, they can have a singular focus without worrying about all the other tasks that executive board members need to do. Having two co-chairs ensures there is variety in the types of activities they plan, and all the work necessary to successfully plan and execute events is covered. An added benefit is that past co-chairs have frequently served in higher positions on the executive board in subsequent years; thus, we frequently have club presidents or vice presidents who have significant experience in planning service events and can give informed guidance to the new co-chairs each year.

A significant task for the community service co-chairs each year is to identify potential community partners and maintain communication and a good relationship with existing partners. Community service co-chairs are encouraged to reach out to existing partners very early in the academic year to ensure that these relationships remain active and productive. The co-chairs also are tasked with proposing new partnerships and are asked to share these ideas with the full board at the opening meeting of each new academic year. This meeting typically occurs before the start of fall classes.

More recently, the club has developed a committee structure allowing for more input from general body members in planning activities. Members of the Community Service Committee are appointed early in the fall semester, and they work with the co-chairs to develop a full year of activities. This is consistently the most popular committee among members, and it is not uncommon that the community service co-chairs start their path to leadership as members of this committee. Additionally, some of the committee members from the greater New Orleans area have introduced the community service co-chairs to new partner organizations.

Funding for extensive outreach programs can always be a challenge, but the Xavier student chapter has developed a strong program for identifying funding for their community service activities. The most significant source of funding has been

ACS student chapter grants. Several years ago, the newly elected board members wrote a successful Community Interaction Student Affiliates (CISA) proposal, and they have continued to successfully secure funding through this program (now known as Community Interaction Grants [CIG]) for the past several years.

The general strategy for securing grant funds has been to have the incoming president and vice president write a grant proposal near the end of the spring semester, then the newly elected community service co-chairs oversee the project the following academic year. As mentioned earlier, these co-chairs often ascend to the positions of president and vice president, so the grant writers each year are generally experienced in overseeing large outreach projects. Some of the recent projects for which the club has received funding include:

- Water Quality in St. Rose, LA: Through this program, club members work with residents of St. Rose, LA, a community that borders a large petrochemical facility. Working with the Louisiana Bucket Brigade, a local environmental action nonprofit organization, club members perform analyses of drinking water samples and communicate the results to residents of the community. The project also includes opportunities for schoolchildren in St. Rose to interact with Xavier students.
- Wow Chemistry Wednesdays: The goal of this project was to provide hands-on laboratory experiences for local high school students. The specific target partners included three local schools at which funding for chemistry laboratory activities was lacking. The central activities of this project have been continued beyond the initial funding period.
- Elements for Those in Elementary: Working with teachers at Ben Franklin Elementary School, club members prepared chemistry demonstrations and hands-on activities for K–5 students. Specific emphasis was placed on students developing their own explanations for what they observed.
- Chemistry in a Box: The goal of this project was to provide local middle school teachers with equipment and supplies for conducting hands-on experiments with students in grades 6–8. Club members organized boxes of all necessary supplies and then visited schools to demonstrate the activities with students. Teachers then kept these supplies in order to provide a similar experience for other students.

More detailed descriptions of certain aspects of these funded projects are provided in the "Middle/High School Outreach Projects" section.

Although the CISA/CIG grants did not require matching funds, the club always included this in its proposals. In earlier years, it would usually match funds through its own treasury and general fund-raising events, but more recently, the group has partnered with different campus programs to secure funding. The Xavier Partnerships for Research and Education in Materials (PREM) program, funded through the National Science Foundation, had outreach activities as a specific component. By partnering with this program, the club was able to secure matching funds for its grants and the PREM program was able to fulfill its outreach requirements. Undergraduate students in these programs and in the club visited one school every week to instruct, tutor, and serve as mentors to elementary school children. The second- and third-grade students were even able to visit Xavier on several occasions during the academic year and during the summers. A similar partnership for securing matching funds currently exists with a large NASA-Minority University Research and Education Project (MUREP) Institutional Research Opportunity (MIRO)-funded initiative on campus.

In writing proposals, students have been encouraged to think about novel ways the club members can work with local K–12 students. One criterion for ACS CIG grants is that the population targeted must include underserved populations. Demographic data show that residents of New Orleans younger than 20 years of age are 68% African American, 3% Asian, and 28% White (3). In addition, the city has seen growth in the local Hispanic population in recent years, with 6% of Hispanic residents being younger than 20 years. With regard to public school enrollment, data from 2014 show that 85% of students are majority African American (4). Science programs in local public schools have been underfunded for years. The Xavier ACS Student Chapter has worked to fill the gaps in science education for these students through the CIG programs they have executed.

In writing grant proposals, the most significant challenge has almost always been developing a budget. A challenge with the budget is predicting the supplies needed and their costs. Although the students have ideas as to which experiments they will perform, final lessons are not planned until consultations with the local teachers are completed. Because we provide the missing labs for underserved schools, we try to budget for unknown items.

Identifying appropriate schools with which to work can sometimes present challenges. However, the Xavier University chapter outreach program in local schools has been able to thrive through building a network of reliable partner schools, while at the same time offering to work with new schools and teachers. The most successful partnerships are those in which teachers and students expect and look forward to visits from Xavier students. Additional details regarding the development of successful partnerships are provided below.

Identifying chapter members who will participate in outreach activities in schools has generally not been a challenge because of the large club membership. What is a challenge, though, is making sure that all members who wish to participate are offered the opportunity to do so. Because the number of schoolchildren in a typical classroom is 20–25, the number of college students who can participate in any specific outreach activity on a particular day is limited. However, by scheduling a large number of such activities, student members can rotate and all who wish to participate are given the opportunity.

Prior to participating in any classroom visit or other outreach activity, chapter members must go through a brief training session. At these sessions, student members are given the opportunity to try the hands-on activities themselves, and they are guided in developing an understanding of the science that is to be taught through these activities. Training sessions are led by a faculty member and a core of student leaders who oversee the outreach projects. A specific effort is made to thoroughly cover all safety aspects of each activity. The club provides all

130

safety personal protection equipment for the volunteers and for the students in the classrooms.

Middle/High School Outreach Projects

The New Orleans urban area has long had a great need for service to provide economically disadvantaged local middle and high school students with a lab experience. Dr. Mehnaaz Ali, a faculty member of Xavier's Chemistry Department, initiated a program in 2012 with the goal of evolving the service project into a service learning class for Xavier students. The program is designed to provide leadership and service opportunities to participating Xavier student volunteers, who develop and teach science experiments within science classrooms of participating schools. The Xavier students lecture on the material and conduct the experiments with the school's students and teachers. In contrast to other outreach initiatives, the 6-week program (per semester) is designed to provide partner schools with a cohesive laboratory component that enhances their science/chemistry curriculum within the classroom. The program has been well-received by middle and high school teachers and students alike, as gauged by comments received from the participants as well as by an anonymous survey conducted at the end of the outreach period. Currently, the program has been successfully implemented within five area high schools and has been recurring every semester since 2012.

Two of the continuing challenges of this program have been to recruit new schools to participate in this collaborative community effort and to retain existing schools within the program. Each Orleans Parish school has a number of bureaucratic steps that need to be followed in order to receive the proper clearance to participate in the program. Dr. Ali has typically formed alliances through word-of-mouth or by setting up meetings with interested science teachers and coordinators for the schools before meeting the administrators. We have seen that if the teachers are invested in participating then it is possible to eventually get the administration to follow suit. Through this mechanism, the program was able to collaborate with seven area schools over a 4-year period.

This process was further aided in 2016 by the funding from the NASA-MIRO grant of which Dr. Ali is a co-principal investigator. One of the grant's major components is to promote an education outreach program with both local schools and 2-year colleges. Dr. Ali, who was responsible for forming alliances with the local middle and high schools, designed a three-day summer teacher workshop at Xavier during the summers of 2016 and 2017. Because the teachers are required to complete a number of professional development hours during the summer, recruiting interested teachers was carried out by mass emailing science teachers from local schools. The inclusion of this program was a great boon to the NASA-MIRO program as well as to the existing teaching outreach effort. The participating teachers were involved in pedagogical teaching seminars presented by Xavier's Center for the Advancement of Teaching and also received hands-on training from Dr. Ali and some NASA-MIRO student volunteers on conducting experiments within their classrooms. They were given experiments and instructions they could bring back to their own classrooms. The program has been a great success as gauged by the anonymous survey conducted at the end of the workshop. The participating teachers have been extremely interested in continuing their collaboration with Xavier and have served as new recruits for the teaching outreach effort. It is expected that some schools will not be retained within the program due to teacher turnover within the schools. Consequently, it is necessary to have a four- or five-school rotation during any given semester for the program to be successful. Thus, the recruitment process is continuously carried out every semester.

Once the program has met the criteria to collaborate with a given school, the next challenge is to work with the teachers and coordinators at the schools to receive their class schedules in a reasonable time frame. Because the teaching outreach program is designed to span 6–8 weeks in a semester, it is necessary to receive the schedule from the school at the beginning of the semester. This ensures the Xavier group has the requisite time to organize meeting times and allows an appropriate number of participating schools to schedule visits for Xavier students. During the program's inception, the Xavier students were recruited from different organizations within the chemistry department as well as from classes Dr. Ali was teaching. However, because the Xavier ACS Chem Club has a requirement for robust participation in service projects (such as outreach) and had already been involved in service projects in local schools for some time, there was a natural synergy to recruit students from the chapter membership. Additionally, since 2015, ACS student chapter students have increased their leadership roles in designing the experiments and organizing the weekly meetings. The team leaders (typically the community service co-chairs) work closely with Dr. Ali to ensure that all the time slots are being filled and that communication with the schools is being carried out efficiently. From experience, we have seen that it is more effective to have a faculty member communicate with the schools (teachers and administrators) with regard to getting cleared for a visit, as well as with determining schedules and topics of interest.

Most of our teachers have been extremely dedicated to the partnership and are eager to share ideas and thoughts regarding possible topics of interest to their classrooms. This information is used to direct the weekly experiments. Additionally, the teachers serve as partners within the classroom to promote an effective learning environment. Our experiences with the classroom teachers have been positive, as gauged by the surveys conducted at the end of the semester.

From the Xavier student's perspective, the teaching outreach provides a fantastic opportunity for leadership and mentoring. The school visits are designed to have the same group of Xavier students visit a particular classroom for the entire semester. The consistent groups ensure the middle and high school students are comfortable with the Xavier students so they are able to initiate a mentor/mentorship role. Additionally, due to the weekly visits, the Xavier students become aware of learning challenges within the classroom or the interests of a particular class and are better able to address the needs of the students to promote a more effective learning environment. The mentor/mentorship roles have worked very well in the past and the Xavier students have requested the same school/classroom for the following spring semester to continue the positive interactions. Since the size of the classes varies from year to year, the number of students assigned to a particular classroom is tailored at the beginning of the semester. Typical class sizes are in the range of 20–25 students. Larger groups typically require at least four to six Xavier students to serve as "leaders" for the classroom. This ensures the students can be divided into smaller groups of four or five with an assigned group leader. For each experiment, one Xavier volunteer serves as the "lecturer" and will describe the concepts of the particular experiment. The Xavier students take turns as the lecturer to ensure that all group members have an opportunity to serve in a leadership role within the classroom.

The teaching outreach program was developed to have each group of Xavier students design a particular experiment that all the schools can implement in a certain week. This requires that the first exploratory meeting for the group at Xavier must occur at least two weeks ahead of the school visits. The weekly meeting times are chosen to accommodate all the Xavier volunteers. It is sometimes possible that volunteers are able to help with the preparation sessions but are unable to find a school to visit due to scheduling conflicts within their own classrooms.

During the first week of the school visit, all the students typically carry out an experiment that has been prepared and lectured on during prior semesters. This "sample" experiment provides the first design group with enough time to develop and prepare for an experiment for the second week of the visit. One such sample experiment is a simple acid/base experiment that is an adaptation of an existing general chemistry titration experiment. The experiment involves the use of ammonia and acetic acid to demonstrate concepts such as neutralization, pH changes, use of phenolphthalein as an indicator, and molar volumes. The experiment is carried out with shell vials and is a fantastic experiment to introduce to the students. Additionally, the experiment is one that can be explored with greater theoretical rigor if being carried out with high school students or can be explained/described at the middle school level without losing any of the important concepts.

The program's success has been dependent on building and sustaining a strong rapport with the participating schools. Additionally, because the ACS Chem Club officers have begun to assume a greater leadership role, it has been extremely important for younger members and officers to interact with Dr. Ali during the planning phase of the semester in order to prepare for more responsibility during later years in the program. This strategy has worked successfully during the 2015–2017 years. Additionally, the officers were able to work closely with Dr. Ali to successfully prepare their Community Interaction Grant (2016), as well as the New Activities Grant for testing the water quality in St. Rose (2017). The current officers and members of the teaching outreach group have played an important role in training the future cohort of student volunteers.

One-Day Outreach Events

While the ACS student chapter at Xavier University of Louisiana has a strong focus on all forms of science and community outreach, we also have several yearly standing commitments. Our large events include activities sponsored by the ACS Louisiana Local Section and the Louisiana Children's Museum. Additionally, our students volunteer with the local STEM NOLA organization once a month and for a few weeks during the summer.

Super Science Saturday

Super Science Saturday is our largest annual group volunteer event, with the greatest number of attendees reached. To commence National Chemistry Week each year, the ACS Louisiana Local Section and the Louisiana Children's Museum host this annual event on a Saturday in October. The event lasts 4 hours and caters to children of all ages; however, most of the attendees are between the ages of 4 and 10 and are accompanied by their parents, grandparents, and teachers. The ACS Louisiana Local Section invites all local colleges and universities, along with representatives from the United States Department of Agriculture and a few high schools, to participate in the event.

Each group is asked to contribute at least one hands-on experiment or demonstration for the children, yet the Xavier ACS Chemistry Club prepares and organizes four experiments every year. While this requires more planning and preparation, our students are excited to have many fun activities and often test different experiments beforehand for the event. While our chemistry club provides two or three of the same popular activities each year, we also attempt to incorporate one or two new experiments that relate to the theme of National Chemistry Week. For example, for the 2017 theme of "Chemistry Rocks!" the students showcased the fluorescent properties of various rocks with both short- and long-wave UV lights and explained common fluorescent and light-emitting items. In 2015, the "Chemistry Colors Our World" theme was introduced, with many colors in the lava lamp and balloon skewer activities. For the 2014 theme of "The Sweet Side of Chemistry: Candy", the students demonstrated the color dyes of M&Ms dissipating in water and explained the science behind the demonstration.

During the week leading up to Super Science Saturday, our student volunteers are trained to conduct all the activities. Drs. Candace Lawrence and Ann Privett provide an introduction to each experiment, explain the chemistry behind it, and show the volunteers the activity once. Also stressed are the safety precautions that need to be addressed and the common behaviors children might display when not being observed carefully. Additionally, the upper-level chemistry club members attend the training session to provide other ideas, experiences, and tricks they have encountered or completed while performing each experiment. After this instruction has been given, the chemistry club members test every activity that will be showcased at Super Science Saturday while they are closely observed. Some of the students gravitate toward certain experiments naturally, and others present new suggestions and ideas to discuss with the children at the event.

Along with learning new experiments and conducting community outreach, the chemistry club members learn how to interact with younger children of all ages. Because each child attendee has a different level of understanding, our student volunteers develop various communication skills to teach each individual. Parents, grandparents, and teachers often accompany the child attendees, so our student volunteers answer a variety of questions on how to conduct the experiments and address any safety concerns. This opportunity significantly improves and expands our chemistry club members' personal and professional interactions.

Because this activity is one of the first major large-scale outreach events of each academic year and is also the most popular annual event, it provides a great opportunity for first-year students and new club members to become more involved. For most of the newer members, this event is their first introduction to science outreach. Because of the nearly 40 volunteers our organization provides for this event, our club members have the opportunity to meet older club members. By working together and teaching children science, the student volunteers are able to make new friendships and acclimate to the university mission and the chemistry club environment. Because we also have a well-established mentor–mentee program, this is a great opportunity for our newer club members to participate in this event with their upper-level mentors.

Our most popular activity, which is conducted at all our large outreach events, is Alka-Seltzer rockets. For this hands-on activity, we use old film canisters, Alka-Seltzer, water, plastic cups, and specially cut PVC tubes. Because this is a favorite activity for so many participants, we train and prepare all students on this event. Beforehand, the students split many Alka-Seltzer tablets in half and fill many film canisters halfway with warm water. All the PVC tubes are placed in plastic cups. Once the preparation is complete, we ask all our participating attendees to wear the safety goggles that we provide. Each child helps by placing the Alka-Seltzer tablet in the film canister. Our student volunteers help the children cap the film canister and flip it upside down into the PVC tube. We allow the children to hold the PVC tube and the cup and aim it toward a poster bull's-eye. Most of our student volunteers also provide ample entertainment and a countdown during the wait time for the rocket.

During this activity, our student volunteers discuss with the children how Alka-Seltzer is used normally and also discuss solubility, the production of carbonic acid, and the subsequent evolution of carbon dioxide gas. At the adjacent lava lamp experiment, the children are introduced to solubility and the non-miscibility of oil and water. Children pick their favorite color, and we allow them to drop the Alka-Seltzer tablet into the lava lamp bottle. While the tablet effervesces, the chemistry club members talk to the children about the science behind the experiment and ask the children questions. As with all our experiments, the club members teach the children all the science at each experiment station and explain how the experiments make use of common household items with which they are familiar. Other examples of experiments that the club has performed are bouncing bubbles, balloon skewer magic, nail polish art, and various forms of slime and putty.

As a large-scale event with at least 300 attendees and typically 30–40 Xavier Chem Club students, this event gives our club members ample opportunity to learn how to communicate scientific ideas and prepares them for discussions in their STEM courses. Most important, our club members are gaining insight into educating the general public about basic principles of science and how they impact our world. This event is also a good bonding and motivation event for the chemistry club members, because they must work together to organize and complete the day of events.

Seuss-I-Cal STEM Day

This is a new event organized by the ACS Louisiana Local Section and the Louisiana Children's Museum in honor of Dr. Seuss's birthday in March. The setup of the event is similar to that of Super Science Saturday, and about 20–30 chemistry club members attend the event. For this event in the past, we have performed our usual popular Alka-Seltzer rockets and lava lamps, but we also added our green slime activity to represent the Dr. Seuss *Green Eggs and Ham* story.

STEM NOLA

On the second Saturday of each month, the STEM NOLA program, under the direction of Dr. Calvin Mackie, sponsors STEM Saturdays at local New Orleans Recreation and Development Commission parks (*5*). Since the inception of STEM Saturdays in the fall of 2013, the chemistry club has provided volunteers, demonstrations, and ideas for the program. For each STEM Saturday, the Xavier faculty liaison, Dr. Florastina Payton-Stewart, organizes the student volunteers and provides training for them on all the experiments. During each event the chemistry club members are also able to network with the other volunteers, who are local professionals and chemists.

Unlike our other outreach activities, STEM NOLA is designed to showcase a single STEM topic with a variety of experiments for all age ranges from grades K–12. At least 10 (and frequently closer to 20) of our chemistry club members attend each month and a few of our club members attend the summer programs. For each month, Dr. Payton-Stewart identifies, organizes, and helps train the student volunteers. Because each STEM Saturday showcases a different subject, the student members have to learn all the activities for the K–12 children. Each age group completes predominantly different activities, but a few experiments are common to all age groups. During the four-hour event, each student volunteer works with the same group of children and helps guide them through each experiment.

NOBCChE Conference

When local and national organizations plan conferences in New Orleans, our chemistry club is often asked to provide volunteers and conduct demonstrations. During the fall of 2014, the National Organization for the Professional Advancement of Black Chemists and Chemical Engineers (NOBCChE) requested that our chemistry club members participate in the STEM Festival. During the four-hour event, our chapter provided four activities for children in the K–12 age range. Because this was also an engineering event, the club members demonstrated other activities, such as memory metal to learn how temperature affects metal coils. Club members also introduced magnetic liquids and ferrofluids and discussed their various properties and applications.

In addition to the STEM Festival event, several of our students volunteered as Quiz Bowl judges and timekeepers. The Quiz Bowl competition was a daylong event for high school students from the entire country. Our students, who participated in the STEM Festival and remained as volunteers throughout the conference, were able to network with many engineers and chemists from all over the country. This was an added benefit to our students, who were able to learn about new research and different career paths.

General Community Service

In addition to the science- and chemistry-related outreach activities already described, the club members participate in numerous other community service activities throughout the year. The strength of this arm of their overall service program has similarly been built on identifying reliable community partners and sustaining relationships. Our community partners have come to expect our contributions each year, and club members, likewise, expect these activities to be part of the overall program of events for the organization each year.

For many years, the organization had a strong partnership with Green Light New Orleans (GLNO). Through this organization, club members worked in targeted neighborhoods to install energy efficient light bulbs in homes. This particular activity was popular with club members for a number of reasons. The nature of the activity allowed for large numbers of club members to participate, and the work was always completed on Saturdays. This was frequently an activity that took place early in the academic year, so club members enjoyed the opportunity to bond as a group outside the classroom and away from campus. They were also highly invested in the goals of GLNO. The opportunity to interact with local residents (and, in some cases, residents of rural areas a bit of a distance from New Orleans) was attractive to club members. They always returned to campus with stories to share about the people they met. Because the work of GLNO was ongoing, the club continued to schedule service Saturdays with them throughout the academic year.

Another more recent partnership has been with the New Orleans Ronald McDonald House Charities of South Louisiana. The Ronald McDonald House serves as a home away from home for families with children who are undergoing long-term care in local New Orleans hospitals. Chemistry club members visit the home twice a semester to cook dinner for and interact with the residents. All supplies for the meal are purchased using funds raised by the club, and all planning and cooking of the meal is coordinated by the Community Service co-chairs. Because this is a recurring and popular event with limited participation, each club member is allowed to participate only once during the year so that the opportunity is made available to all interested club members.

Club officers also plan several large single-day community service events each year. One of the more successful recent events was a fund-raiser for childhood cancer research that brought together several campus organizations and students for a Saturday kickball tournament. The Kick-It-For-Cancer organization is a national nonprofit group that helps other groups organize these events. The event itself was a one-day kickball tournament held on campus, and all campus organizations were afforded the opportunity to enter a team.

The ACS Chem Club has twice recently been recognized as the Xavier campus Organization of the Year (2014–2015 and 2016–2017) and was given special campus recognition for Outstanding Service in 2013 and 2016. These accolades are primarily due to the visibility of the group on campus as a leader in providing service. Annually, club members are visible on campus through sponsoring Mass during National Chemistry Week, handing out Halloween candy to local children as part of the Trick-a-Trunk program, and through participation in the United Negro College Fund Walk. They have conducted numerous successful food, toiletry, and clothing drives and have regularly participated in events and activities sponsored by other campus organizations (e.g., Agrowtopia, a local urban gardening initiative). Critical to success in this area are the strong connections of club officers to our Student Government Association (SGA) and Interorganizational Council (IOC). Officers regularly attend IOC and SGA meetings and include reports from these organizations on the agenda of each general body meeting of the club. Through the club, members are continually kept abreast of service opportunities on campus and each year officers continue to set as goals maintaining and strengthening relationships with other campus entities.

Conclusion

While the Xavier ACS Student Chapter sponsors activities in all areas expected by ACS (e.g., professional development, chapter development, social activities, fund-raising) the overall community service program has been central to the success of the chapter for many years. In 2016–2017 alone, the chapter participated in more than 30 such events, and club members have come to recognize the importance of these activities. Through the organization of the executive board, mentoring of new members, and long-range planning, the club has been able to sustain this successful program of service for nearly 20 years. Due to the dedication of club members, officers, and faculty to the mission of the university and the goals of the organization, there can be no doubt that this strong record of serving the community will continue to be the defining characteristic of the Xavier ACS Student Chapter for many years to come.

Acknowledgments

Many have contributed to the success of the Xavier ACS Chem Club program of service over the years. We wish to thank all past and present club officers for their dedication to the mission of the organization. In addition, we are grateful for the contributions of Dr. Florastina Payton-Stewart in fostering the relationship of the club with the STEM NOLA program. We are thankful for the ongoing support we receive from the ACS Louisiana Local Section. Finally, we are most appreciative of the continued financial support we have received through New

Activities Grant (NAG) and CIG grants offered through the ACS Undergraduate Programs Office.

References

1. Tischler, J. L.; Wilhelm, C. A.; Wilhelm, M. R. Building a Stronger Chapter and Richer Community through Science Outreach Partnerships. In *Building and Maintaining Award-Winning ACS Student Member Chapters Volume 2: Specific Program Areas*; Mio, M. J., Benvenuto, M. A., Eds.; ACS Symposium Series 1230; American Chemical Society: Washington, DC, 2016; pp 53–65.
2. Adams, M. The Xavier University of Louisiana Student ACS Chapter: An Organization Guided by a University Mission. In *Building and Maintaining Award-Winning ACS Student Member Chapters Volume 1: Holistic Viewpoints*; Mio, M. J., Benvenuto, M. A., Eds.; ACS Symposium Series 1229; American Chemical Society: Washington, DC, 2016; pp 123–136.
3. Perry, A. The New Orleans Youth Index 2016. https://www.datacenterresearch.org/reports_analysis/the-new-orleans-youth-index-2016/ (accessed Dec. 13, 2017).
4. Jacobs, L. New Orleans by the Numbers: Public School Enrollment. https://educatenow.net/2015/01/28/new-orleans-by-the-numbers-public-school-enrollment/ (accessed Dec. 13, 2017).
5. STEM NOLA. http://www.stemnola.com (accessed Dec. 28, 2017).

Editors' Biographies

Matthew J. Mio

Matthew Mio is a Professor at the University of Detroit Mercy in the Department of Chemistry and Biochemistry. His research focuses on new transition metal catalyzed cross-coupling reactions. Projects include exploring both the mechanism and synthetic capabilities of these reactions, with particular emphasis on the generation of phenylacetylenes for use in nanoelectronics and supramolecular chemistry. He is also interested in studying the pedagogy of organic chemistry. He has been co-advisor to the Detroit Mercy Chemistry Club (American Chemical Society Student Members) for over 16 years.

Mio holds a B.S. in chemistry from the University of Detroit Mercy and a Ph.D. in organic chemistry from the University of Illinois at Urbana-Champaign. He was awarded a Mellon Fellowship to perform post-doctoral research and teaching at Macalester College (St. Paul, MN). Mio joined Detroit Mercy's faculty in 2002.

Mark A. Benvenuto

Mark Benvenuto is a Professor of Chemistry at the University of Detroit Mercy, in the Department of Chemistry and Biochemistry. His research spans a wide array of subjects, including the use of energy-dispersive X-ray fluorescence spectroscopy to determine trace metal elements in land-based and aquatic plant matter, especially with regard to use in phyto-remediation of soil.

Benvenuto received a B.S. in chemistry from the Virginia Military Institute, and after several years in the US Army, a Ph.D. in inorganic chemistry from the University of Virginia. After a post-doctoral fellowship at the Pennsylvania State University, he joined the faculty at the University of Detroit Mercy in 1993.

Indexes

Author Index

Subject Index

147

Printed in the USA/Agawam, MA
May 1, 2019

702036.001